T0212735

Lecture Notes in Computer Science   **8433**

Commenced Publication in 1973
Founding and Former Series Editors:
Gerhard Goos, Juris Hartmanis, and Jan van Leeuwen

More information about this series at http://www.springer.com/series/7409

Theodor Wyeld · Paul Calder
Haifeng Shen (Eds.)

Computer-Human Interaction

Cognitive Effects of Spatial Interaction, Learning, and Ability

25th Australian Computer-Human Interaction Conference, OzCHI 2013
Adelaide, SA, Australia, November 25–29, 2013
Revised and Extended Papers

Editors
Theodor Wyeld
Flinders University
Bedford Park, SA
Australia

Paul Calder
Flinders University
Bedford Park, SA
Australia

Haifeng Shen
Flinders University
Bedford Park, SA
Australia

ISSN 0302-9743   ISSN 1611-3349 (electronic)
Lecture Notes in Computer Science
ISBN 978-3-319-16939-2   ISBN 978-3-319-16940-8   (eBook)
DOI 10.1007/978-3-319-16940-8

Library of Congress Control Number: 2015937403

LNCS Sublibrary: SL3 – Informations Systems and Applications, incl. Internet/Web, and HCI

Springer International Publishing AG Switzerland is part of Springer Science+Business Media
(www.springer.com)

Preface

This book is a direct outgrowth of OzCHI 2013, the annual conference of the Computer-Human Interaction Special Interest Group (CHISIG) of the Human Factors and Ergonomics Society of Australia (HFSA) and Australia's leading forum for a growing international community of practitioners, researchers, academics, and students to exchange work in all areas of Human-Computer Interaction.

The conference Technical Program Committee comprised of 142 researchers, among whom 77 were from Australia and 65 were from overseas. The committee received submissions for 71 long papers, 83 short papers, and 38 student design challenge entries, from countries in Asia-Pacific, Europe, North Europe, North America, and South America. After a rigorous peer-review process, they accepted 34 long papers, 45 short papers, and 9 student design challenge finalists, overall 46% of submissions. All long and short papers were subject to double-blind peer-review with each long paper reviewed by at least three committee members and each short paper reviewed by at least two committee members. The conference proceedings were published by CHISIG and appear in the Association for Computing Machinery (ACM) Digital Library (dl.acm.org).

The theme of the conference was Augmentation, Application, Innovation, and Collaboration, which reflects a variety of technical and social challenges in designing and deploying human-centred computer applications through augmenting our daily lives with innovative interaction and collaboration technologies. The program covered a wide range of topics around this theme, including "Ubiquitous Computing," "Interface, Interaction, and Visualization," "Health and Welfare," "Learning Environments," "Gaming," "Mobile and Touch Interaction," and "Social and Collaboration Technologies."

Following the conference, 19 papers were invited as extended full chapters that addressed issues related to the cognitive effects of spatial interaction, learning, and ability. Of these 11 were accepted after a two-stage double-blind peer-review process with at least three reviewers for each paper. The chapters in this book are the result.

November 2013   Paul Calder

Introduction to the Book

This book is about the cognitive effects of spatial interaction, learning, and ability. Eleven chapters that are significant extensions to the conference papers published at the OzCHI 2013 proceedings were selected. From these, four core themes emerge: *multi-dimensional interaction, video gaming, spatial learning,* and *physical spatial interaction.* Each of these themes is addressed by several chapters. Collectively, they contemporaneously address these themes within an HCI context.

The section on *multi-dimensional interaction* addresses spatial interaction that involves 3D, 4D, and touch-based environments. This section consists of four chapters. Bradley Wesson and Brett Wilkinson investigate the use of full body gestures to facilitate 3D sculpting through a natural user interface providing users with the ability to sculpt a virtual clay-like substance into different forms. They also evaluated, perceived, and measured speed and accuracy, as well as the users reported fatigue and ease of use through a secondary interface augmenting the application with a separate set of gestures on a handheld smartphone, while supplementing the positional data provided by the Kinect with orientation data. Takanobu Miwa et al. then present an interactive 4D visualization technique to control a 4D viewing direction via handling of principal vanishing points. They propose an algorithm utilizing principal vanishing points as an interface for the intuitive 4D viewing direction control and apply it to a system that enables users to fly through a 4D space. Ashley Colley et al. report perceptual, cognitive, and behavioral aspects revealed by their user studies that consider stereoscopic 3D touch screens of different sizes in static and mobile usage contexts. They also present the requirements for interacting with stereoscopic 3D touch screen user interfaces. Seongkook Heo presents design strategies for rich-touch interactions and applications they developed with a novel touchpad prototype, capable of measuring a finger hover as well as finger force applied to the screen.

The section on video gaming addresses spatial perception, spatial ability, and cognitive complexity in video games. This section consists of three chapters. Tuukka M. Takala et al. investigate the use of volumetric shadows for enhancing 3D spatial perception and action in third-person motion games. Their studies indicate that different volumetric shadow cues affect both user experience and gameplay performance positively or negatively, depending on the lighting setup. Playing was found to be most easy and fluent in a typical virtual reality setting with stereo rendering and flat surface shadows. Theodor Wyeld and Benedict Williams describe a study of gamers and non-gamers and their spatial ability to draw what they see. The results show that gamers tended to perform better than non-gamers in both spatial reasoning tests and drawing tasks. Jemma Harris et al. examine information acquisition behavior of pilots in response to a series of simulated flight sequences involving different levels of cognitive complexity. Their results suggest that assessments of the complexity of a task should be employed as a benchmark in task assessment.

The section on spatial learning addresses how spatial interaction can affect a student's learning ability. Winyu Chinthammit et al. developed a basic 3D molecule construction simulation to augment the teaching of organic chemistry by helping students grasp the concepts of chemistry through visualization in an immersive environment, 3D natural interaction, and audio feedback. Their results show that immersive interactive systems have great potential to promote visuo-spatial reasoning by externalizing complex spatial relationships and providing interactive tools for direct manipulation of spatial objects. Mark Reilly et al. present the interface design of a mobile real-time collaborative note-taking application that can allow a small self-selecting group of students to proactively keep each other engaged while still allowing for individuals to contribute and interact, with regard to their own ability and preferred learning style. A pivotal design goal is to accommodate students' diverse cognitive abilities by spatially separating their workspaces and allowing each individual to choose the most suitable way to interact with their peers' workspaces. Their results show that students are more engaged in a lecture with collaborative rather than individual note-taking, and are more satisfied with sharing spatially separated workspaces than a common workspace.

The section of physical spatial interaction addresses the physical aspects of spatial interaction. Mahmood Ashraf and Masitah Ghazali propose a physicality-focused quantitative evaluation method to assist embedded system developers in managing the interaction complexities of their products and then conduct a user study to assess the acceptance of the embedded system developers toward the proposed method. They further propose a range of values for appliances where these values can be treated as a catalog or as guidelines as they design and develop physical interfaces and again evaluate the acceptance of embedded system developers toward the proposed catalog. Henrik Sørensen et al. investigate proxemic interaction, specifically the potentials and challenges of spatial interaction spanning across separate physical locations. They develop a multi-room music system that extends Apple AirPlay to allow spatial interaction with one's music player and perform a field evaluation of the system. The results reveal a number of findings related to the cognitive perception of the spaces it is used in, such as importance of a simple interaction, the importance of providing local interaction, the challenge of foreground and background interactions, and challenges in designing interaction with music in discrete zones.

Theodor Wyeld
Paul Calder
Haifeng Shen

List of Technical Program Committee Members

Safurah Abdul Jalil
Elin Eliana Abdul Rahim
Truna Aka J.Turner
Ons Al-Shamaileh
Leena Arhippainen
Magnus Bang
Debjanee Barua
Mark Billinghurst
David Brown
George Buchanan
Paris Buttfield-Addison
Marina Buzzi
Maria Claudia Buzzi
Paul Calder
Linda Candy
Siyuan Chen
Aaron Chen
Caslon Chua
Andrew Clayphan
Nathalie Colineau
Karin Coninx
Sally Jo Cunningham
Xianghua Ding
Matthew D'Orazio
Claire Dormann
Wendy Doube
John Downs
Andreas Duenser
Kirsten Ellis
Ulrich Engelke
Viv Farrell
Graham Farrell
Eric Fassbender
Zac Fitz-Walter
Marcus Foth
Jill Freyne
Henry Gardner
Martin Gibbs
Voula Gkatzidou
Stephen Green
John Grundy
Florian Güldenpfennig
Karen Henricksen
Luke Hespanhol
Seamus Hickey

Xavier Ho
Zaana Howard
Andrew Johnston
Syahrul Junaini
Doris Jung
Rohit Ashok Khot
Khamsum Kinley
Lone Koefoed Hansen
Henrik Korsgaard
Lars Kulik
Tobias Langlotz
Geehyuk Lee
Hyowon Lee
Tuck Wah Leong
Chun-Cheng Lin
Christopher Lueg
Martin Luerssen
Christof Lutteroth
Peter Lyle
John Manning
Evi Indriasari Mansor
Roberto Martinez-
Maldonado
Kevin McGee
Dana Mckay
Richard Medland
Alejandra Mery Keitel
Seyed Hadi Mirisaee
Johannes Mueller
Florian Nachreiner
Hideyuki Nakanishi
Suranga Nanayakkara
Bjorn Nansen
Erik G. Nilsson
Susanna Nilsson
Tim Nugent
Piia Nurkka
Kenton O'Hara
David O'Hare
Jeni Paay
Helen Partridge
Jon Pearce
Sonja Pedell
Abdul Moiz Penkar
Bernd Ploderer

Vesna Popovic
Peter Purgathofer
Kenneth Radke
Patrick Rau
Fiona Redhead
Toni Robertson
Christine Satchell
Jennifer Seevinck
Haifeng Shen
Hirohito Shibata
Simeon Simoff
Petr Slovak
Ross Smith
Wei Song
Fabius Steinberger
Duncan Stevenson
Ozge Subasi
Susanne Tak
Sampo Teräs
Jimmy Ti
Feng Tian
Claire Timpany
Helena Tobiasson
Martin Tomitsch
Keith Unsworth
Keith Vander Linden
Nicholas Vanderschantz
Frank Vetere
Greg Wadley
Tony Wang
Jenny Waycott
Gerald Weber
Michael Weber
Ian Welch
Christoph Wimmer
Brett Wilkinson
Nicholas Wittison
Chui Yin Wong
Clinton Woodward
Burkhard Wuensche
Theodor Wyeld
Lonce Wyse
Hiroaki Yano

Contents

Physical Spatial Interaction

Multi-dimensional Interaction

Evaluating 3D Sculpting Through Natural User Interfaces Across Multiple Devices

Bradley Wesson$^{(\boxtimes)}$ and Brett Wilkinson

Computer Science and Engineering, Flinders University, Adelaide, Australia
{brad.wesson,brett.wilkinson}@flinders.edu.au

Abstract. Traditional digital sculpting software offers a wealth of specific-purpose tools discouraging artist play in the environment. These tools interact with the projection of the geometry to a flat screen, through the use of a mouse or graphics tablet - devices poorly suited to this domain given their lack of a direct method for indicating the depth of interactions. This project investigated the use of full body gestures to facilitate such artistic expression. Skeletal data drives a natural user interface providing users with the ability to sculpt a virtual clay-like substance into different forms. The application was tested in this mode, as well as with a supporting secondary interface for perceived and measured speed and accuracy, users' reported fatigue and ease of use. This secondary interface provided various touch gestures on a smartphone held in the user's right hand, while supplementing the positional data provided by the Kinect with orientation data. Results indicated that users were able learn the interface quickly, but depth-perception, grip detection and speech performance were lacking. The secondary interface resulted in fewer undo events, though users reported it as offering little benefit and awkward to use.

Keywords: Natural user interfaces · Artistic expression · Kinect · Smartphone

1 Introduction

Current 3D modelling packages can be cumbersome to use for the design and creation of organic forms, primarily because of the complexity of multi-tool software [39] and the inability to support ambiguous, imprecise, and incremental interactions [13]. These complex interfaces are ill-suited to the creative process [36] and typically drive designers to prototype ideas with pencil and paper before attempting to realise their ideas in a 3D environment. Interactions with these tools take place through the projection of the geometry onto a 2D viewport, and require the use of a 2D pointing device, such as a mouse or graphics tablet. As such, interacting with the third dimension requires viewport rotations to swap one of the viewport axes for the third into-the-screen axis. Mesh occlusion in this arrangement causes selection of inner faces to become tedious and awkward; some environments attempt to overcome this problem by cycling through occluded faces upon multiple clicks but this can be a slow and frustrating experience.

Motion control systems, such as Microsoft's Kinect, offer a unique way to support natural user interface gestures through body movements and voice commands, which

© Springer International Publishing Switzerland 2015
T. Wyeld et al. (Eds.): OzCHI 2013, LNCS 8433, pp. 3–20, 2015.
DOI: 10.1007/978-3-319-16940-8_1

may offer a solution to the above problems. As interactions with these sensors take place in all three spatial dimensions, 3D interactions become much simpler and more natural. Technical limitations impose some restrictions on the types of interactions currently possible; although some work has gone into capturing individual finger articulations, this data remains expensive to capture and offers unstable results [29]. This leads most Kinect applications to rely on easier to capture full-body gestures with more distinct poses and movements, resulting in more physical exertion.

Complementing the Kinect skeletal data with an orientation sensing mobile phone offers a significant increase in data available to the system. The compass, gyroscope, and accelerometer provide acceleration and rotation data, combined with the positional data captured by the Kinect, allows the application to provide a virtual wand in the environment anchored and rotating with the user's hand. With 65 % of Australian adults already owning a mobile smartphone [7], this additional technology requirement is expected to have a negligible impact on the availability and affordability of access to the system.

This project investigated this pairing of devices to facilitate interaction with a virtual scene. With the 3D data captured through skeletal tracking, users can easily, naturally, and immediately give depth to their interactions. This method of input is not possible through traditional 2D input devices, which rely on viewport rotations or use of the scroll wheel to indicate depth [40]. Orientation sensing provided by the mobile phone offers a greater capability to react to subtle wrist movements, with the touch, vibration, and physical buttons providing further interaction interfaces.

An application was developed to test the usefulness of these interactions for creative work and the impact that the secondary mobile smartphone has. The application provided users with a virtual clay-like object and a single grab-and-pull gesture to alter the objects shape. A marching cubes variant was used for display, with various visual effects aimed at providing an enhanced sense of depth.

Analysis of user session recordings show that users were able to learn to use the interface within a matter of minutes, with median session durations of 130 S for the initial teapot, and 75 S for the second smiley face model. Asymmetrical teapots indicate that test participants had trouble perceiving the depth cues provided by the system – particularly misinterpreting the glow as volumetric. Participants reported the jittery tracking and unreliability of speech recognition as common problems, as well as seeing little benefit in the addition of the assisting mobile phone.

2   Related Work

Creative thought is believed to be a complex process of drawing on ideas in unique and interesting ways; Poincaré has been quoted observing that "to create consists of making new combinations of associative elements which are useful" [27]. This primary-process thought is most effectively performed with the unconscious mind [26], excelling in inattentive thought [41]. The discovery that results offers many emotional benefits [3, 20].

Designing natural user interfaces and interaction gestures to support the creative process requires thought about how users interact with creative applications to communicate their intent. Donald Schön has outlined a number of key principles of this process. He explains that when confronted by a complex problem, users will attempt to alter the problem until it can be solved with a known pattern. He talks about the see-move-see pattern most users will use with design, and the discovery of unintended consequences in other domains. In this pattern, designers will interpret the current state of an object, perform an action to alter the object, and observe the changes made to influence their next action [34].

These interactions are often vague, imprecise, and ambiguous. Current solutions, such as CAD software, require precise and exact interactions, and discourage users from playing in the environment [13]. This harms the creative process, which relies on progressively augmenting the form, fixing errors, or refining detail. This process allows users to create far more elaborate forms that would otherwise be tedious or impossible [28]. Other applications such as Mudbox and ZBrush allow users to play in a virtual clay-like environment with less precise interactions, but currently offer only 2D input interactions and impose unintuitive structural restrictions.

Other studies observe that the increase in non-traditional input devices such as 3D pointing devices and hand gestural systems have revealed the trouble many users have with understanding 3D spaces – that many cues available in the physical world cannot be replicated in a virtual space. Novel metaphors must be developed to better communicate this missing information and increase the users understanding of the space. [5].

A number of researchers have previously attempted to design natural interactions for generating 3D models through the use of sketching gestures, such as ILoveSketch [2] and EasyToy [23]. FiberMesh and TEDDY provide variations on these ideas, producing implicit surfaces based on drawn curves [18].

While most of these solutions rely on 2D interfaces such as a mouse or graphics tablet, other projects have attempted to investigate the suitability of interacting through 3D interfaces, most notably through the Playstation Move controller and the Microsoft Kinect sensor bar. The 3D Puppetry animation tool [15] and the interactive blocks in DuploTrack [14] make use of the Kinect sensor to capture interactions with real world objects and replicate them in the virtual environment, allowing intuitive movements along all three axes. As they describe, these applications allow for users to interact with the virtual space through natural play and performances, to undertake otherwise complex tasks.

Alternatively, researchers have used this approach to guide the creation of real world clay sculpts, by projecting hints onto the physical clay based on this depth data [32]. This technique allows for organic modelling with the addition of touch sensation from working with a physical medium. Taking this a step further, others have attempted to augment the physical to virtual mapping to provide haptic feedback for virtual interactions [21] or to overcome physical limitations [10].

The use of body gestures in a learning environment has the potential to ignite creativity [16], and may be beneficial for improving motor control in disabled individuals [8, 17]. However, some studies suggest that this mode of input is slower and less accurate than traditional interfaces; likely due to a lack of user familiarity and unrefined hardware [33].

The pairing of an RGB camera and depth sensor in the Kinect, along with software to track the movements of the scene, can allow automatic model generation.

KinectFusion [19] and other projects have used this data to virtually reconstruct real-world scenes and objects by simply rotating the object within the Kinect viewport.

Using multiple input devices together to offer greater contextual information about the gestures invoked by the user can provide interfaces that are more efficient, accurate, and flexible [30]. This increase in flexibility allows the interface to support a wider range of users, tasks, and environments; the user can choose the interaction mode that best suits them or their situation, and switch between modes as their situation changes.

Conveniently, low-cost devices sporting capacitive multi-touch-optimised interfaces have recently become far more prevalent [4]. A set of "de"; facto standard touch gestures has since been developed to interact with virtual objects [38]; some popular gestures include the pinch-to-zoom gesture for enlarging items, and swipe gestures to transition between views and categories. The specifications race that has taken place since this burst of activity has seen the introduction of many integrated sensors; the average smartphone contains at least an accelerometer and compass [22], enabling it to orient itself to Earth's gravitational and magnetic field. These sensors are commonly used to orient the phones screen to remain aligned with gravity, and to allow navigation maps to be presented correctly. With the addition of a gyroscope, this combination of sensors offers the ability for the device to provide quickly-responding orientation and positioning data through sensor fusion [35].

3 Rationale

The Kinect sensor was developed to support whole-body gestures for the Xbox 360 gaming console to broaden the range of gamers able to access the system [11]. The sensor is able to track the positions of major joints in the human skeleton, and estimate orientations, but these readings can be quite inaccurate. Having such inaccurate and high latency feedback prevents use of subtle movements, causing use of the Kinect sensor to typically result in physical exertion. This can be beneficial to health [12], but prolonged use can result in fatigue [25], not a desirable trait for creative applications.

The pairing of a mobile device in the users hand may mitigate this problem by providing greater data capturing capabilities for subtle movements and providing a simpler means of gathering contextual information through button presses and touch interactions.

Recent updates to the Kinect SDK have offered the ability to detect hand states (open or closed), through "grip" detection, offering cleaner support for bimodal interactions. In addition, the 4 microphone array provided by the sensor allows capturing and recognition of verbal commands, giving applications the potential for a simple and elegant method of invoking commands and mode changes.

4 Application Design

Our primary goal when designing an application experimenting with natural user interactions through 3D body gestures was to offer familiar interactions in an intuitive manner. Users interact with an infinite solid blob through push and pull motions. The user reforms the initially spherical shape by moving their hand to the desired location, clenching their fist or pressing a physical button to 'grab' the geometry at that location, and then drag it in another direction. This allows a single bimodal gesture to be used to extrude or indent the geometry. A slider provides the ability to adjust the interaction radius, allowing this same gesture to be used for large form changes and minor detailing. Figure 1 shows the hand tool, slider for controlling the brush size and example teapot.

Fig. 1. Application features and completed teapot

Often the user will require viewing or zooming in on a different part of the model. For this, the application provides users with a reorientation gesture, triggered by placing both hands in front of the user's body, 'gripping' the empty space, and moving their hands in a coordinated manner. The user is able to pull their arms outward or together to scale the model, pivot their hands about each other to rotate, and move their hands in unison to translate the model. While an assisting smartphone is connected, the user can instead perform these gestures on the smartphone's touchscreen by placing two fingers on the screen surface and similarly moving their fingers about the screen. While performing these gestures, movements of the phone are interpreted as model orientation commands, so that rotating the phone body while interacting with the touchscreen caused the model to rotate likewise.

The aim of these gestures is to always provide immediate feedback to the user and increase the usefulness of interactions; ensuring transparency and usefulness of actions invoked can improve user acceptance of an application and therefore engagement. Providing users with this feedback can dramatically increase the user's confidence in the system [1]. This feedback allows the user to better learn the workings and limitations of the application and better accommodate for its flaws, reducing frustration from misinterpretations of user intent. Actions undertaken by the application should be as useful as possible in order to reduce the number of required tools presented to the user, and increase the perceived usefulness.

In addition to these gestures, the application provides 3 verbal commands. These can be activated at any time by speaking aloud to the system, and provide a means of returning to a known state. The smartphone would emit a short vibration pulse after successful activation, to confirm to the user that the command has been registered.

- The *"Reset Camera"* command instructs the system to revert the models orientation to its initial state. This is useful if a user becomes disoriented and wishes to return to a known state. This instruction causes the camera to interpolate between its previous orientation and the default orientation, so that the user can gain a sense of where they were prior to invoking the action.

- The *"Undo"* command offers the ability to remove the most recent change to the models geometry. This allows the user to be more careless with their interactions and experiment with the system, and is a very important feature for supporting creative thinking [31].
- The *"Reset Scene"* command resets the system back to its initial state, replacing the current model with the default sphere shape. This is invoked when a user has completed their task and wishes to start the next task, or if they become stuck and wish to start again.

5 Implementation

An application was developed using the XNA graphics framework, implementing the gestures described previously. A voxel structure stored the model data, which was converted to a triangulated mesh before rendering. A Kinect sensor tracked the user's skeleton with some smoothing applied to reduce jitter, and provided a microphone array for capturing voice.

The application made use of an infinite grid of voxels to store model data; voxels were grouped into "chunks" for efficiency, which are allocated as the user extends the model into new regions. Each voxel stored a density value to aid in calculating a smoother model surface. The surface was defined as the shell at which the density passes 0.5, and thus can lie anywhere between two voxel centres.

Conversion of this structure to polygons for rendering followed a slight variation of the marching cubes algorithm [24]. These edges were stored in a flat array following the order shown in Fig. 2. The red lines are iterated in such a way as to ensure that the enumerator remains consistently inside or outside of the solid, until an edge vertex is found or until the system has determined that the surface does not penetrate this cube. The order of enumeration starts at the origin, and traverses through edges [1, 3, 4, 7, 10–12]. This minimum set of 7 edges ensures that all intersection configurations will be found.

Any edges found with vertices crossing the surface density threshold trigger a second integrator. This iterator iterates clockwise around the edge loop defined as

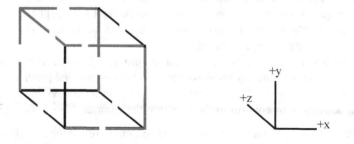

Fig. 2. Ordering of edge vertices

bounding the face to the right of the intersection. This process halts and is repeated with any other edges found to cross the surface density threshold until the original edge is found (physical constraints require that each edge loop has an even number of intersections, and surface intersections will form a closed loop). The resulting series of intersections describe the polygonal representation of the bounding voxel data and were then triangulated for rendering.

The grab and pull gesture was implemented as a 3D analogue of the 2D smudge tool in photo editing software, working on the 3D voxel grid instead of a 2D image. This tool pulls values in the voxel grid along the direction of interaction. The natural blurring that results from this interaction aids in removing surface noise, though if left uncorrected could lead to the interaction behaviour changing over time. Based on whether the interaction began inside or outside of the model, a correction factor was added or removed from the surrounding voxels, pushing their values toward the maximum or minimum.

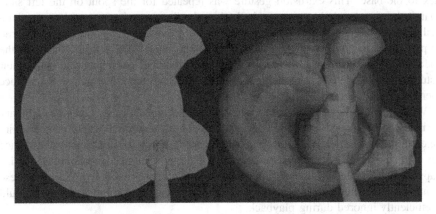

Fig. 3. Model without shading (left) and with shading (right)

Shading the surface of the model was important to communicate as much information about the structure of the model as possible, Fig. 3 indicates the effect of shading on the model work space. The model was lit with ambient light creating shadows in crevices, and a subtle blue light emanating from the interaction point, and a more powerful red light denoted the interaction region. This ambient occlusion used the fact that the density values representing the model were somewhat blurred, and was able to sample the 6 neighbouring voxels, averaging the density, to determine if the current voxel was in a crevice (where neighbouring voxels would typically contain quite high densities) or in the open (where densities would be lower).

6   Evaluation

Preliminary testing of the application by 13 users revealed some common behaviours and considerations for further refinement. Participant interactions were recorded and observed, and each participant completed a questionnaire after completion. Verbal feedback was noted as they used the system.

Participants were taken to a room with the application running on a 15" laptop. The Kinect sensor was positioned above the laptop, looking over the screen toward the user. Participants stood about 3 m from the laptop while interacting. In some tests, participants would be required to hold a 4.5" Nokia smartphone in their right hand.

Participants were required to make use of the tools available in the application to transform a sphere into two predefined 3D forms. These forms were selected for their commonality and ability to test a wide range of uses: extruding; joining; flattening; cutting; and poking. The participants were required to repeat these tests twice, once using the Kinect on its own and once again using the Kinect with an assisting mobile phone. The order that the participant performs the tests under each of these input configurations was alternated, to reduce the impact this ordering has on results.

For the teapot, participants were required to extrude a handle out of the right side of the model, beginning their gesture beneath the surface, pulling outward, and joining back to the base. This extrusion gesture was repeated for the spout on the left side. A larger brush was used to reshape the body to be flatter on the top and bottom and a smaller brush to create a handle for the teapot lid. Producing the smiley face required the participant to punch two eye holes by beginning their gesture from outside of the model and pushing inward. Producing the mouth required participants to begin their gesture from outside of the model, carve out a hole, and then drag across the surface where the mouth should be.

While performing each of these tests, the system logged all user actions to a binary file ready to be played back for later analysis. These recordings captured the entire skeleton at constant intervals, the state of the user's hands and optionally the device they were holding, when the user pressed a button or invoked a command, and a snapshot of the effected chunks after each interaction was recorded. Each of these events was time stamped, and the file was structured such that unimportant parts could be efficiently ignored during playback.

The population of users involved in these tests consisted primarily of Flinders University CSEM (Computer Science, Engineering and Mathematics) students. Of those involved, 69 % had experience using the Nintendo Wii motion controller, 54 % had previously used Kinect interactives, and 34 % had used the PlayStation Move controller. 15 % had never used a gaming motion controller before (Kinect, Move, WiiMote, or otherwise).

7 Results

As stated above, the test interactions were repeated once with the assisting smartphone and once without, with the order alternated between tests. The log of the application state allowed for future playback and data analysis. From this data, the number of brush strokes, undo events and duration of testing sessions could be extracted and compared, as well as qualitative measures, such as the final model smoothness and quality. The following charts in Figs. 4 and 5 show the number of brush strokes and undo events observed from the recorded sessions.

A similar number of brush strokes was observed with and without the assisting mobile phone, however the number of undo events was reduced (p-value = 0.0863).

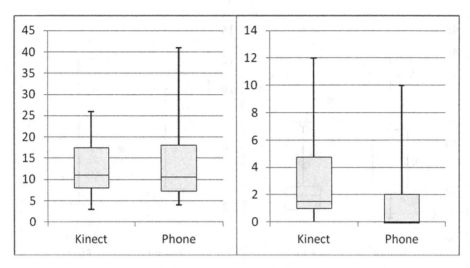

Fig. 4. Number of brush strokes   **Fig. 5.** Number of undo events

The limited number of participants involved in the trial prevents conclusive results; however this seems to indicate that participants made fewer mistakes with the assisting mobile smartphone. An average of 1.62 undo events were issued while making use of the assisting smartphone, compared to 3.08 for tasks with just the Kinect.

All participants were tasked with constructing the teapot model before the smiley face for both modes of the test (with and without the mobile smartphone). Although the teapot only had two required features, the spout and the handle, the complexity of those parts required continual refinement to properly shape. Because of this, creating the teapot required more operations than the smiley face. Interestingly, although all participants already had considerable experience with the system and required fewer brush strokes to produce the face, the number of undo commands used remained relatively similar, with two participants using as many as 12 undo commands for this task. Figures 6 and 7 display the number of brush strokes and undo events observed for these two tasks.

Figure 8 shows the durations for various tasks, from the first event until the last. The teapot required more time to complete, averaging 130 S to complete, while the smiley face took just 75 S on average. This measurement may correspond to increased familiarity of the system gestures after completion of the teapot. Further analysis indicates that people spent longer with an assisting smartphone when creating the Teapot model (p-value = 0.022). The results, although inconclusive, also indicate slightly quicker completion when creating the smiley face model with an assisting smartphone (p-value = 0.299).

After completing tests, participants were required to fill out a survey capturing information about their experience with the application. This information provides an insight into how they felt about how the two modes compare.

Participants were asked to rank how fast they felt they would be able to use the application in each mode after becoming familiar with the interface. Their responses are

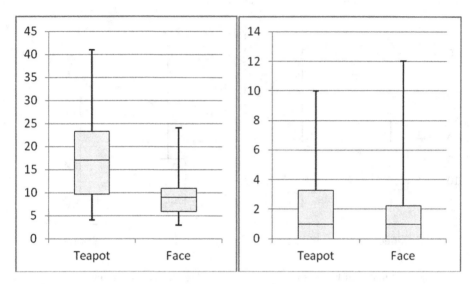

Fig. 6. Number of brush strokes   **Fig. 7.** Number of undo events

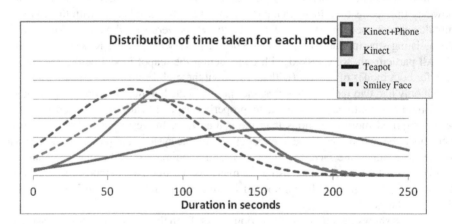

Fig. 8. Task duration comparison

visualised in Fig. 9; ranging from 0 (slowest) to 10 (fastest). This comparison indicates that most users felt that the Kinect on its own offered a faster experience than when coupled with the phone (p-value = 0.028).

Participants then answered which mode they felt allowed them to produce a higher quality model. They answered this on an 11 point scale between the Kinect on its own and the Kinect with an assisting mobile phone. These results, visualised in Fig. 10, indicate a slight preference to using the system without the assisting mobile phone to produce higher quality models, although the results were fairly inconclusive, with many participants considering the two to be quite similar.

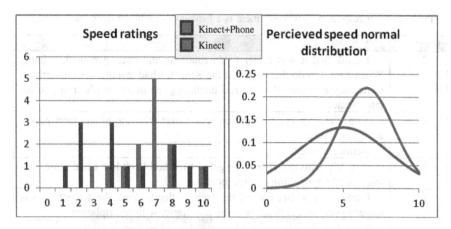

Fig. 9. Speed ratings given by participants

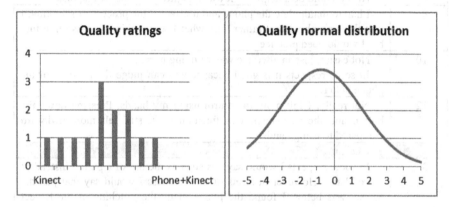

Fig. 10. Quality ratings given by participants

These results indicate that the addition of the mobile device did not offer greater perceived speed or quality for creation of models. When asked if participants felt the addition of a smartphone provided added benefit, the responses were divided. The following Table 1 displays the 5 positive responses, 5 negative responses, and 3 unclear responses. Of the positive responses, three participants outlined the increase in control afforded by the use of a physical button, over the less-responsive 'gripping' Kinect gesture. Other participants, such as User 12, mentioned that this action was unnatural and caused them to feel less in control. An interesting remark from user 1 was the recognition that the rotation of the device requires the arm to bend to odd angles, as alluded to by Zhai with flying mice implementations [40]. A number of participants have also mentioned the inaccuracy and jittery performance that the Kinect experienced while holding the mobile phone.

Participants were asked which features of the application were most frustrating to use, these results are available in Table 2. In their response, three participants

Table 1. User responses to whether there is a benefit of combined interaction devices

User	Yes
2	I think that it was easier to do rotations and scaling with the phone. When I tried to do this with just the Kinect, I had trouble with sometimes switching to 'drawing' mode and damaging the model while trying to rotate / scale.
4	Yup, the gripping recognition was easier with the phone than purely using the Kinect
5	Better for fine tuning
11	Yes. My arms didn't seem to get as tired and I felt like I had more control over the modelling environment.
13	Currently yes, because of the lack of depth using the Kinect. It was easier to control the depth because familiarity with smartphones
	No
1	No, I had to bend my arm at odd angles to position the hand
8	No - it could be a helpful if it was less jittery and better response.
9	I had to manipulate the phone and it affected the process of sculpting, I was thinking about tool rather than what I was making. It was quite fun, but I would need practice.
10	Not easier, just an alternate way of doing it. In some respects it is good because you can judge the position of the phone better.
12	Not really, I felt more in control using my hands. Pressing down the button and the movement was functional but still felt more awkward compared to using hands.
	Unclear
3	In some respects I would say yes such as the use of a button instead of closed hand. However in terms of intuitiveness I would say that just the Kinect was better. I found the phone made the glitching of the Kinect less... :)
6	For complex tasks the phone was difficult to do careful work. When drawing the smiley face however the phone was much easier to use.
7	The phone assisted with some of the precision for drawing when not requiring a rotation during the smiley face task. It was difficult to combine rotation with the extrusion though when trying to draw a tea pot.

mentioned problems with the voice recognition, particularly with the commonly used undo command. Another three users mentioned the jittering of data captured by the Kinect, and while using the assisting phone, causing errors in their model. Other mentions of note were the request for a redo feature, the inability to judge depth, and the unresponsive 'grab' gesture detection by the Kinect.

Participants consistently produced teapots with a bias toward the camera; 9 out of 13 participants would ensure the red glow was visible before beginning their action. This glow indicates the part of the surface that will be influenced by any actions taken. The consequence of users beginning their action at this point is that the features of the

Table 2. Features reported by the user as frustrating

User	Response
1	Trying to drag the smile of the face up and along in a line at the same time.
2	The voice worked okay some of the time, but certainly wasn't reliable. That was the most annoying when I needed to undo a catastrophic mistake and it wouldn't work.
3	The random Kinect positioning glitches
4	There is no re-do!
5	Nothing
6	The complex testing with the phone was most frustrating as I did not feel I had as much control on the movement and aspect of the model
7	The voice command for "undo" did not respond well probably due to the echo in the room. Trying to rotate with the phone was difficult. And the jitter when your hand was over the body with the Kinect caused increased error rates.
8	Phone jitter
9	Being tested on something the first time I used it. Rotating the hand on the phone was tricky. Need strong hands to press the buttons.
10	Just making accidents.
11	Kinect voice recognition.
12	Accidentally grabbing hold when my hand wasn't fully open.
13	Depth

teapots were consistently drawn closer to the viewer and not along the teapots plane of symmetry as they should have been. Figure 11 provides an example of participant's work with the system. The top right element of the image shows the top down view of the model they created. Such a situation may be avoidable in future trials by having this glow represent the interaction volume.

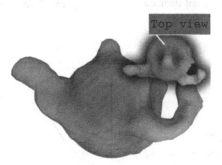

Fig. 11. A teapot created by one of the participants

Many participants had trouble producing the mouth of the smiley face. They would attempt to start the gesture on the surface, expecting the application to recognise that they intended to carve out the area brushed, however this action is ambiguous, offering the application no indication of whether it should cut into or pull out of the model. The application instead attempts to retain the surface shape from the initial gesture position, resulting in little change. Many participants would also attempt carving out the mouth with the default brush size, only to undo and repeat their action with a more appropriate brush size; the recorded data indicated that the majority of undo events were applied to the mouth.

Participants also mentioned trouble with the application initiating gestures while their hand was still open; typically occurring while their hand was parallel to the camera, or too close to their body. They also struggled with invoking verbal commands, often needing to repeat their command multiple times before it was recognised, and sometimes having the application invoke the wrong command (particularly troublesome with destructive commands, such as resetting the scene).

After completing their feedback about their use of the application, participants were asked to rank the severity of a number of different metrics in the two application modes, on a scale from 0 to 10. The results were fairly similar, however there is a slight indication that users felt that the Kinect with an assisting mobile phone offered less physical demand (Fig. 13), but required more mental effort (Fig. 12) to properly use.

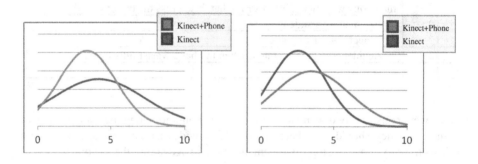

Fig. 12. Perceived mental demand   Fig. 13. Perceived physical demand

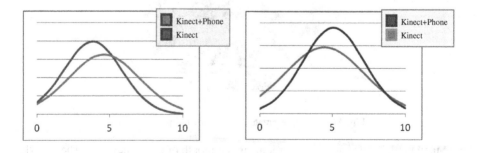

Fig. 14. Perceived frustration comparison   Fig. 15. Perceived effort comparison

Users were also asked to rate each environment with relation to the effort they required to create their model, and the frustration experienced after using the system in each mode. These results showed very little difference in the perception of users for either mode. Figures 14 and 15 indicate the comparative perceptions of users.

From the above data we can infer that the additional interaction device was not perceived as an encumbrance by the users. This user perception combined with the fine grained control and reduced error that occurred with the mobile smartphone assisted development indicates that for rapid prototyping this style of combined input could enhance the experience.

8 Future Work

The skeletal tracking provided by the Kinect is naturally improving with the introduction of upgraded hardware and SDK updates, offering new functionality and improving tracking accuracy. Complimenting this data with an additional hand-held device capable of detecting hand articulation and motion can lessen the reliance on unreliable Kinect functionality. A sensor fusion process could use the data capture from the smartphone to calculate gravity-independent device acceleration [35], which can be matched against the Kinect to provide more responsive movement detection.

The additional sensors in the smartphone could be exploited in other ways. Most smartphones contain multiple high-quality microphones for noise suppression and speakerphone functionality. This microphone is also held closer to the user's mouth during application use; using it over the Kinect microphone may provide better voice isolation and speech detection. The high-quality touchscreen and processor of these devices is capable of rendering and displaying detailed information; this auxiliary screen could be used to show another perspective of the model and further information about the application state.

While a smartphone is adequate for this application, data-gloves may improve experience, due to their increased sensor and feedback capabilities. The newly developed ControlVR is a data-glove capable of detecting full arm movements through the use of inertial sensors placed in a glove and set of wearable bands [6]. The Myo armband is capable of similar data capturing for hand articulations with a lower cost and less intrusive setup [37], though would likely need to be used in combination with a Kinect for full arm tracking. Du et al. [9] has investigated the use of force feedback in these data-gloves; this could serve as a way for the user to "feel" when they are touching the model surface.

Initial testing revealed some trouble with the provided surface-based interaction glow, as mentioned earlier, it is likely that altering this to indicate glow within the solid would aid in depth perception. The addition of new depth cues (e.g. more advanced shadowing), haptic feedback (e.g. data gloves), or 3D-capable video hardware (e.g. Oculus Rift) could also aid in delivering sufficient information to the user. The primary concern when offering depth perception cues to the user is to provide a quick and efficient method for the user to determine the location in which their interaction with the model will take place.

9 Conclusion

This project investigated the use of multiple input interfaces, specifically a Kinect sensor and an orientation sensing smartphone, to facilitate an intuitive natural user interface for shaping a blobby virtual material into various forms. An application was developed and tested by 13 participants, who were tasked with producing a teapot and smiley face model with the provided tools.

These tests revealed that while users were able to learn and understand the gestures required quickly (completing each model in 1 to 2 min), many had trouble perceiving depth with the provided cues, resulting in teapots being created off-centre. Grip gesture recognition was troublesome under certain conditions, causing erroneous activation of the grab tool, though this problem was lessened by the addition of an undo capability. Vocal commands proved to be unreliable at times, but complimented the gesture-based controls well, providing users with a simple method of controlling application state, without hindering their use of gestures.

Participants were conflicted on whether they felt the addition of a smartphone improved the experience, some noting that it gave them more control over the application while others mentioned that it was awkward to use and offered them little benefit. Analysis of the recorded sessions was likewise inconclusive; similar durations were observed, though a decrease in the number of undo events was observed with the addition of the mobile device.

References

1. Antifakos, S., Kern, N., Schiele, B., Schwaninger, A.: Towards improving trust in context-aware systems by displaying system confidence. In: Proceedings of the 7th International Conference on Human Computer Interaction with Mobile Devices & Services, pp. 9–14. Salzburg, ACM (2005)
2. Bae, S.-H., Balakrishnan, R., Singh, K.: ILoveSketch: as-natural-as-possible sketching system for creating 3d curve models. In: UIST 2008 Proceedings of the 21st Annual ACM Symposium on User Interface Software and Technology, pp. 151–160. ACM, New York (2008)
3. Banker, J.E.: Family clay sculpting. J. Fam. Psychother. **19**(3), 291–297 (2008). doi:10.1080/08975350802269533
4. Barrett, G., Omote, R.: Projected-capacitive touch technology. Inf. Disp. **26**(3), 16–21 (2010)
5. Bowman, D.A., Kruijff, E., Laviola, J.J., Poupyrev, I.: An introduction to 3D user interface design. In: Presence: Teleoperators and Virtual Environments, pp. 96–108 (2001)
6. Control, VR.: The future of Virtual Reality, Animation and more, ControlVR. http://www.controlvr.com/ (2014). Accessed (6 June 2014)
7. Deepend: Australian mobile device ownership and home usage report 2014. Sydney (2014)
8. Diment, L., Hobbs, D., Chau, T.: A gesture-based virtual art program for children with severe motor impairments: development and pilot study. In: Proceedings of the 7th International Convention on Rehabilitation Engineering and Assistive Technology, p. 65. START, Singapore (2013)

9. Du, H., Xiong, W., Wang, Z., Chen, L.: Design of a new type of pneumatic force feedback data glove. Fluid Power and Mechatronics (FPM). In: 2011 International Conference, pp. 292–296 (2011). doi:10.1109/FPM.2011.6045775

10. Engel, D., Curio, C., Tcheang, L., Mohler, B., Bülthoff, H.H.: A psychophysically calibrated controller for navigating through large environments in a limited free-walking space. In: Proceedings of the 2008 ACM Symposium on Virtual Reality Software and Technology, pp. 157–164. ACM, Bordeaux (2008). doi:10.1145/1450579.1450612

11. Fitzgibbon, A., Tsunoda, K.: Kinect for Xbox360. (Microsoft, Performer) Microsoft Research Faculty Summit, Redmond, Washington, USA (2010). http://research.microsoft. com/en-us/um/redmond/events/fs2010/presentations/Kudo_Fitzgibbons_Kinect_for_ Xbox360_071210_FacSummit.pdf. Accessed 12 July 2010

12. Ganesan, S., Anthony, L.: Using the kinect to encourage older adults to exercise: a prototype. In: Proceedings of the 2012 ACM Annual Conference Extended Abstracts On Human Factors in Computing Systems Extended Abstracts, pp. 2297–2302. ACM, New York (2012)

13. Gross, M.D.: Ambiguous Intentions: a Paper-like Interface for Creative Design, pp. 183–192. ACM (1996)

14. Gupta, A.: Closing the loop between intentions and actions. In: Adjunct Proceedings of the 25th Annual ACM Symposium on User Interface Software and Technology, pp. 27–30. ACM, New York (2012)

15. Held, R., Gupta, A., Curless, B., Agrawala, M.: 3D puppetry: a kinect-based interface for 3D animation. In: Proceedings of the 25th Annual ACM Symposium on User Interface Software and Technology, pp. 423–434. ACM, New York (2012)

16. Hsu, H.-M.J.: The potential of kinect in education. Int. J. Inf. Educ. Technol. 1, 365–370 (2011)

17. Huang, J.-D.: Kinerehab: a kinect-based system for physical rehabilitation: a pilot study for young adults with motor disabilities. In: The Proceedings of the 13th International ACM SIGACCESS Conference on Computers and Accessibility, pp. 319–320. ACM, New York (2011)

18. Igarashi, T., Matsuoka, S., Tanaka, H.: Teddy: a sketching interface for 3D freeform design. In: ACM SIGGRAPH 2007 Courses. ACM, San Diego (2007). http://doi.acm.org/10.1145/ 1281500.1281532 (Accessed)

19. Izadi, S., Kim, D., Hilliges, O., Molyneaux, D., Newcombe, R., Kohli, P., Fitzgibbon, A.: KinectFusion: real-time 3D reconstruction and interaction using a moving depth camera. In: Proceedings of the 24th Annual ACM Symposium on User Interface Software and Technology, pp. 559–568. ACM, New York (2011)

20. Kimport, E.R., Robbins, S.J.: Efficacy of creative clay work for reducing negative mood: a randomized controlled trial. Art Ther. 29(2), 74–79 (2012). doi:10.1080/07421656.2012. 680048

21. Kohli, L.: Warping virtual space for low-cost haptic feedback. In: Proceedings of the ACM SIGGRAPH Symposium on Interactive 3D Graphics and Games, pp. 195–195 (2013). ACM, Orlando. doi:10.1145/2448196.2448243

22. Lane, N., Miluzzo, E., Lu, H., Peebles, D., Choudhury, T., Campbell, A.: A survey of mobile phone sensing. IEEE Commun. Mag. 48(9), 140–150 (2010). doi:10.1109/MCOM. 2010.5560598

23. Liu, Jin, Y.-J.: EasyToy: plush toy design using editable sketching curves. In: IEEE Computer Graphics and Applications, pp. 49–57. IEEE, Beijing (2011)

24. Lorensen, W.E., Cline, H.E.: Marching cubes: a high resolution 3D surface construction algorithm. In: SIGGRAPH 1987 Proceedings of the 14th Annual Conference on Computer Graphics and Interactive Techniques, pp. 163–169. ACM, New York (1987)

25. Lu, Z., Chen, L., Fan, C., Chen, G.: Physiological signals based fatigue prediction model for motion sensing games. In: Nijholt, A., Romão, T., Reidsma, D. (eds.) ACE 2012. LNCS, vol. 7624, pp. 533–536. Springer, Heidelberg (2012)
26. Martindale, C., Rindos, D.: On Hedonic Selection, Random Variation, and the Direction of Cultural Evolution. Current Anthropology **27**(1), 50–53. (February) http://www.jstor.org/stable/2743028
27. Mednick, S.A.: The associative basis of the creative process. Psychol. Rev. **69**, 220–232 (1962)
28. Nealen, A., Igarashi, T., Sorkine, O., Alexa, M.: FiberMesh: designing freeform surfaces with 3D curves. ACM Trans. Graph. **26**, 41 (2007)
29. Oikonomidis, I., Kyriazis, N., Argyros, A.: Efficient model-based 3D tracking of hand articulations using kinect. In: Proceedings of the British Machine Vision Conference, pp. 101.1–101.11. BMVA Press (2011)
30. Oviatt, S.: Ten myths of multimodal interaction. Commun. ACM **42**(11), 74–81 (1999). doi:10.1145/319382.319398
31. Resnick, M., Myers, B., Nakakoji, K., Shneiderman, B., Pausch, R., Selker, T., Eisenberg, M.: Design Principles for Tools to Support Creative Thinking (2005)
32. Rivers, A., Adams, A., Durand, F.: Sculpting by numbers. ACM Trans. Graph. **31**(6), 1–157 (2012). doi:10.1145/2366145.2366176
33. Sambrooks, L., Wilkinson, B.: Comparison of gestural, touch, and mouse interaction with fitts' law. In: Proceedings of the 25th Australian Computer-Human Interaction Conference: Augmentation, Application, Innovation, Collaboration, pp. 119–122. ACM, Adelaide, Australia (2013)
34. Schön, D.: Designing as reflective conversation with the materials of a design situation. In: Artificial Intelligence in Design, pp. 3–14. Department of Urban Studies & Planning, Cambridge (1992)
35. Shala, U.: Indoor Positioning using Sensor-fusion in Android Devices. Kristianstad University, School of Health and Society. Kristianstad: Kristianstad University (2011). http://urn.kb.se/resolve?urn=urn:nbn:se:hkr:diva-8869 (Accessed)
36. Sheng, J., Balakrishnan, R., Singh, K. (n.d.). An Interface for Virtual 3D Sculpting via Physical Proxy
37. Thalmic Labs.: THALMICLABS. (2014). https://www.thalmic.com/en/myo/ Accessed Myo - Gesture control armband by Thalmic Labs
38. Villamor, C., Willis, D., Wroblewski, L.: Touch Gesture Reference Guide. (2010, April 15). LukeW Ideation + Design: http://static.lukew.com/TouchGestureGuide.pdf. Accessed 22 June 2013
39. Zeleznik, R.C., Herndon, K.P., Hughes, J.F.: SKETCH: an interface for sketching 3D scenes. In: ACM SIGGRAPH 2006 Courses. ACM, Boston (2006)
40. Zhai, S.: User performance in relation to 3D input device design. SIGGRAPH Comput. Graph. **32**(4), 50–54 (1998). doi:10.1145/307710.307728
41. Zhong, C.-B., Dijksterhuis, A., Galinsky, A.D.: The merits of unconscious thought in creativity. Psychol. Sci. **19**(9), 918–921 (2008)

Interactive Four-dimensional Space Exploration Using Viewing Direction Control Based on Principal Vanishing Points Operation

Takanobu Miwa[1]([⊠]), Yukihito Sakai[2], and Shuji Hashimoto[3]

[1] Graduate School of Advanced Science and Engineering, Waseda University,
3-4-1 Okubo, Shinjuku-ku, Tokyo 169-8555, Japan
takmiwa@shalab.phys.waseda.ac.jp
http://takanobumiwa.wordpress.com/
[2] Faculty of Engineering, Fukuoka University, 8-19-1 Nanakuma, Jonan-ku,
Fukuoka 814-0180, Japan
yukihito@fukuoka-u.ac.jp
[3] Faculty of Science and Engineering, Waseda University,
3-4-1 Okubo, Shinjuku-ku, Tokyo 169-8555, Japan
shuji@waseda.jp
http://www.shalab.phys.waseda.ac.jp/

Abstract. This chapter presents an interactive 4-D visualization technique that controls a 4-D viewing direction via handling of principal vanishing points. Principal vanishing points are represented by projecting points at infinity of 4-D principal coordinate axes onto 3-D space. Our previous studies have confirmed that, because the principal vanishing points relate to a 4-D eye-point and the 4-D viewing direction, they can be landmarks to intuitively move in 4-D space. In this chapter, we propose an algorithm that utilizes principal vanishing points as an interface for the intuitive 4-D viewing direction control, and apply an algorithm to a framework of a system which enables one to fly through 4-D space. The developed system can achieve to visualize and explore an intricate 4-D scene such as a maze in 4-D space. We evaluate effectiveness of the proposed 4-D interaction technique by user experiments.

Keywords: 4-D space · 4-D viewing direction · 4-D interaction · 4-D interface · 4-D visualization · Principal vanishing points

1 Introduction

A concept of "dimension" forms a basis for human knowledge, science and culture. Although it is said that the humans live in the physical world of three dimensions, some mathematicians, thinkers and artists have realized that they can imagine 4-D space beyond 3-D space without any problems. Here, 4-D space is space defined by four axes which are orthogonal to each other. However, in general, 4-D space has been mysterious and incomprehensible for the humans, because the humans have only experience of 3-D space and cannot see 4-D space directly with their eyes. Conversely, if an interactive environment which can allow the user to look around and move in 4-D space is provided,

© Springer International Publishing Switzerland 2015
T. Wyeld et al. (Eds.): OzCHI 2013, LNCS 8433, pp. 21–46, 2015.
DOI: 10.1007/978-3-319-16940-8_2

it is expected that humans will be able to obtain intuitive understanding of 4-D space. In fact, there exists research reporting that mathematicians can suddenly feel 4-D space while they interact with a computer visualized 4-D object [5].

This chapter presents a novel interactive 4-D visualization technique that controls a viewing direction in 4-D space via a pick and move operation of principal vanishing points in 3-D space. The principal vanishing points are represented by projecting points at infinity of 4-D principal coordinate axes onto 3-D space. From our previous studies, we found that principal vanishing points can be landmarks to intuitively move in 4-D space, because they closely correlate with a 4-D eye-point and the 4-D viewing direction. We utilize the principal vanishing points as an interface for the intuitive 4-D viewing direction control. Our proposed algorithm can determine the 4-D viewing direction from the positions of the principal vanishing points in 3-D space. Furthermore, we apply the algorithm to a framework of a system which enables one to fly through 4-D space. The developed system consists of recent commodity components such as a personal computer, a motion sensor, a head mounted display with a built in a 6-DoF sensor and a five-button wireless mouse. Using the system, a user can freely explore 4-D space with a 4-D fly-through action, while smoothly changing the 4-D viewing direction by the simple handling of principal vanishing points. Through user experiments, we confirm usability and efficiency of the proposed 4-D interaction technique.

This chapter is organized as follows. Firstly, related work and our previous work on 4-D visualization and interaction are introduced in Sect. 2. A relationship between the 4-D viewing direction and the principal vanishing points is shown in Sect. 3. In Sect. 4, we propose the algorithm to determine the viewing direction in 4-D space from the principal vanishing points in 3-D space. Then, we outline the interactive system that enables the user to intuitively explore 4-D space in Sect. 5. In Sect. 6, we demonstrate some 4-D scenes such as insides of 4-D solids, multiple solids and a 4-D maze using the system. In Sect. 7, we discuss effectiveness of our 4-D interaction technique by user experiments. Finally, conclusions are given in Sect. 8.

This chapter is a revised and extended version of our paper entitled "Four-dimensional Viewing Direction Control by Principal Vanishing Points Operation and Its Application to Four-dimensional Fly-through Experience" [13].

2 Related Work and Our Previous Work

There are a number of approaches to visualizing multi-dimensional data via dimensionality reduction techniques. A principal components analysis, a multi-dimensional scaling and a parallel coordinate plot are typical methods. These methods are useful to analyze multi-dimensional statistical data. However, they are not suited to overview higher-dimensional geometric data without missing any information.

Starting with the pioneering work of Abbott [1] and Banchoff [3], visualization of the 4-D objects has been studied in various fields. Most of 4-D visualization techniques project the 4-D object into 3-D space, based on an analogy of that we can imagine a 3-D object from its 2-D projection drawings [6, 8, 10, 11, 14]. There are some work that slices 4-D data with a hyperplane in 4-D space [15, 22], by extending in such a way that we cut the 3-D object with a 2-D plane. As an extension of 3-D computer graphics techniques such as a lighting model and GPU computing, shading and lighting techniques in 4-D space have also been studied [4, 9, 21]. As mentioned above, various rendering methods of 4-D objects have been reported. Here, we focus our attention on research using the projection method for the 4-D object [6, 8, 10, 11, 14]. An advantage of projecting the 4-D object to 3-D space from different eye-points in 4-D space is that it keeps various original 4-D geometric features, not only a structural continuity but also parallelism and orthogonality. In approaches to 4-D visualization, this method will be effective to observe and understand shapes of the 4-D object. However, in some work [6, 8, 10], the 4-D eye-point is fixed. In the others [11, 14], changes of the 4-D eye-point and a 4-D viewing direction are limited. Therefore, it is not sufficient to move freely around 4-D space and overview the 4-D object in the same manner of 3-D space.

Some researchers have attempted to develop 4-D interaction techniques by associating common input devices such as a mouse, keyboard input, a joystick and a touch-screen with geometric operations of the 4-D object [2, 7, 23]. For example, these approaches allow a user to interactively observe rotations of the 4-D object. However, the user cannot move in 4-D space, because of the fixed 4-D eye-point. As described above, although some research into approaches to 4-D interaction has been conducted, direct manipulation interfaces to freely explore 4-D space have not been developed so far.

In our former work, we constructed a 4-D visualization algorithm and a 4-D geometric algorithm via 5-D homogeneous processing [16, 18, 19]. 5-D homogeneous processing is uniformly expressed using 5×5 matrices and 5×5 determinants. Different from the conventional visualization techniques [6, 8, 10, 11, 14] and the interaction techniques [2, 7, 23], our algorithm enables us to visualize various 3-D perspective drawings of any 4-D data onto 3-D space from an arbitrary 4-D eye-point, viewing direction and viewing field. Moreover, 5-D homogeneous processing can be applied to all points including points at infinity in 4-D space. That allows the display of principal vanishing points overlaying the 3-D perspective drawings of 4-D data, by projecting the points at infinity in directions of 4-D principal coordinate axes onto 3-D space.

By extending this 4-D visualization algorithm, we developed an intuitive and interactive 4-D space display system that made human actions in 3-D space correspond to movements and moving directions of the eye-point in 4-D space [16]. By controlling a flight-controller pad associated with human actions, the user can observe various types of 4-D data such as a 4-D solid, 3-D time-series data and 4-D mathematical data with an arbitrary 4-D viewing field, while moving around a 4-D spherical surface which surrounds the observed 4-D data [17, 18, 20]. Through our studies, we found that users utilized principal vanishing points as a landmark to understand their own location in 4-D space, when they moved around 4-D space. Inspired by this perception, we developed a novel intuitive and interactive system that employed the principal vanishing points as an interface to control the 4-D eye-point movement [12]. Using the system, we moved the 4-D eye-point along the 4-D spherical surface by simply and directly moving the principal vanishing points in 3-D space. From results of user experiments, we showed that the system using the principal vanishing points operation [12] can provide more intuitive and interactive 4-D interaction than the former system using the flight-controller pad associated with human actions [16].

Following on from our latest study [12], in this chapter, we extend the principal vanishing points operation to the 4-D viewing direction control. Furthermore, we apply this framework to a 4-D fly-through action which enables us to freely travel around 4-D space.

3   Relationship Between 4-D Viewing Direction and Principal Vanishing Points

In this section, we describe a relationship between the viewing direction in 4-D space and the principal vanishing points in 3-D space. We obtain 3-D perspective drawings of 4-D data by converting data defined in the 4-D world-coordinate system $x_w y_w z_w w_w$ those of 3-D screen-coordinate system $x_s y_s z_s w_s$ [16, 18]. Figure 1 shows the 4-D visualization model to observe a 4-D solid from an arbitrary eye-point, viewing direction and distance in 4-D space. The 4-D viewing direction is defined as the direction from the 4-D eye-point p_f $(x_{p_f}, y_{p_f}, z_{p_f}, w_{p_f})$ to the 4-D observed point p_a $(x_{p_a}, y_{p_a}, z_{p_a}, w_{p_a})$ in the 4-D world-coordinate system, and coincides with the negative direction of the w_e-axis of the 4-D eye-coordinate system $x_e y_e z_e w_e$, the origin of which lies at the 4-D eye-point. The center of the 3-D screen and the center of the background hyperplane are located on the 4-D visual axis at distance h and $f (> h)$ from the 4-D eye-point, respectively. The 3-D screen has a dimension of $2k \times 2k \times 2k$ in the $x_s y_s z_s$-space. The 4-D viewing field is defined as a truncated pyramid that is formed by the 4-D eye-point, the 3-D screen and the background hyperplane. Only 4-D data inside the 4-D viewing field is visualized as the 3-D perspective drawing on the 3-D screen. This visualization algorithm is composed of a view field transformation, a perspective transformation and a clipping operation in 4-D space. Including points at infinity in 4-D space, the transformation from the data of the 4-D world-coordinate system to that of the 3-D screen-coordinate system is represented in homogeneous coordinates V_s as the following equation:

$$
\begin{aligned}
V_s &= [X_s \quad Y_s \quad Z_s \quad W_s \quad v_s] \\
&= [X_w \quad Y_w \quad Z_w \quad W_w \quad v_w] \boldsymbol{T}_v(p_f, p_a) \boldsymbol{T}_p(k, h, f), \\
x_s &= \frac{X_s}{v_s}, \; y_s = \frac{Y_s}{v_s}, \; z_s = \frac{Z_s}{v_s}, \; w_s = \frac{W_s}{v_s},
\end{aligned}
\tag{1}
$$

where the transformation matrices $\boldsymbol{T}_v$ and $\boldsymbol{T}_p$ are the 4-D view field transformation matrix and the 4-D perspective transformation matrix, respectively. This algorithm enables one to observe various types of 4-D data from an arbitrary 4-D viewing direction at an arbitrary 4-D eye-point. Moreover, by changing the parameters k, h and f of the 4-D perspective transformation matrix $\boldsymbol{T}_p$, we can visualize 4-D data not only with various 4-D viewing fields but also with projection methods such as a perspective projection, a parallel projection and a slice operation.

Now, we consider the relationship between the viewing direction in 4-D space and the principal vanishing points in 3-D space. The points at infinity in x_w-, y_w-, z_w- and w_w-directions are represented as $V_{x_w} (1,0,0,0,0), V_{y_w} (0,1,0,0,0), V_{z_w} (0,0,1,0,0),$

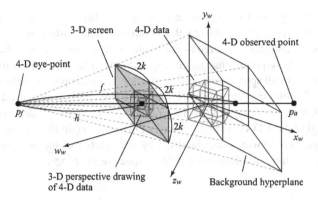

3-D screen   4-D data
4-D eye-point $2k$ 4-D observed point
f $2k$
Pf h pa
$2k$ x_w
w_w
3-D perspective drawing z_w Background hyperplane
of 4-D data

Fig. 1. Visualization model of 4-D space.

and $V_{z_w}\overline{(0,0,0,1,0)}$. Substituting these points to Eq. (1), the principal vanishing points vp_x, vp_y, vp_z and vp_w on the 3-D screen are obtained as follows:

$$
\begin{aligned}
vp_x &= \left(x_{vp_x}, y_{vp_x}, z_{vp_x}\right) \\
&= \left(\frac{h}{k}\frac{1}{\tan\beta\tan\gamma}, -\frac{h}{k}\tan\gamma, 0\right), \\
vp_y &= \left(x_{vp_y}, y_{vp_y}, z_{vp_y}\right) \\
&= \left(0, \frac{h}{k}\frac{1}{\tan\gamma}, 0\right), \\
vp_z &= \left(x_{vp_z}, y_{vp_z}, z_{vp_z}\right) \\
&= \left(-\frac{h\tan\beta}{k\cos\gamma}, -\frac{h}{k}\tan\gamma, -\frac{h}{k}\frac{1}{\tan\alpha\cos\beta\cos\gamma}\right), \\
vp_w &= \left(x_{vp_w}, y_{vp_w}, z_{vp_w}\right) \\
&= \left(-\frac{h\tan\beta}{k\cos\gamma}, -\frac{h}{k}\tan\gamma, \frac{h}{k}\frac{\tan\alpha}{\cos\beta\cos\gamma}\right),
\end{aligned}
\tag{2}
$$

where α, β and γ are the parameters regarding the 4-D viewing direction for the 4-D view field transformation, which are derived from the 4-D eye-point p_f and the 4-D observed point p_a [16, 18]. It is assumed from the converse relation of Eq. (2) that we can calculate the parameters regarding the 4-D viewing direction using some given principal vanishing points. Thus, it is expected that the principal vanishing points will be useful for a geometric reference to detect the viewing direction and the position during moving in 4-D space.

Figure 2 shows the relationship between the principal vanishing points and the 3-D perspective drawings of a hypercube. In each picture shown in Fig. 2, the 3-D screen corresponds to an inside of a cubic region closed by a white dashed line. Corresponding to changes of the 4-D eye-point and the 4-D viewing direction, various 3-D perspective drawings of the hypercube inside the 4-D viewing field appear on the 3-D screen. On the other hand, the points at infinity in x_w-, y_w-, z_w- and w_w-directions are displayed as four principal vanishing points vp_x, vp_y, vp_z and vp_w in the entire 3-D space without defining any 4-D viewing field. According to the geometric relationship of the 4-D world-coordinate system and the 3-D screen-coordinate system, positions and the number of the principal vanishing points change, correspondingly. The principal vanishing points are represented by small spheres with different color. For example, as the top-left picture shown in Fig. 2, although one principal vanishing point vp_w is displayed on the 3-D screen, the other three principal vanishing points vp_x, vp_y and vp_z also appear in 3-D space by changing the 4-D viewing direction as the bottom-right picture shown in Fig. 2. The 3-D perspective drawings are classified by the number of the principal vanishing points; one-, two-, tree- and four-point 3-D perspective drawings.

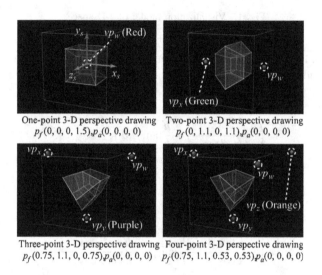

One-point 3-D perspective drawing Two-point 3-D perspective drawing
$p_f(0, 0, 0, 1.5)$,$p_a(0, 0, 0, 0)$ $p_f(0, 1.1, 0, 1.1)$,$p_a(0, 0, 0, 0)$

Three-point 3-D perspective drawing Four-point 3-D perspective drawing
$p_f(0.75, 1.1, 0, 0.75)$,$p_a(0, 0, 0, 0)$ $p_f(0.75, 1.1, 0.53, 0.53)$,$p_a(0, 0, 0, 0)$

Fig. 2. Relationship between the principal vanishing points and the 3-D perspective drawings of a hypercube. The parameters k, h and f regarding the 4-D viewing field are given as 0.5, 0.5 and 100, respectively.

4 Algorithm of 4-D Viewing Direction Control via Principal Vanishing Points Operation

In this section, we explain an algorithm to determine the viewing direction in 4-D space from principal vanishing points in 3-D space. This algorithm is mainly composed of two processing steps. The first one estimates positions of principal vanishing points. The second one estimates parameters regarding the 4-D viewing direction. In order to apply to the 4-D visualization algorithm described in Sect. 3, in the second processing step, the position of the 4-D observed point is calculated from the parameters regarding the 4-D viewing direction. Different from our previous 4-D visualization method [12], we can freely look in all 4-D viewing directions from an arbitrary 4-D eye-point via simple handling of the principal vanishing points.

Let us discuss the first processing step. When one principal vanishing point is picked and moved in 3-D space, the other three principal vanishing points will be automatically allocated at the corresponding correct positions so that they satisfy their geometric positional relationship in 3-D space. Suppose, for instance, the 3-D perspective drawing of 4-D data and principal vanishing points vp_{x_b}, vp_{y_b}, vp_{z_b} and vp_{w_b} are displayed in 3-D space to steer a visual axis in the direction from the 4-D eye-point p_f to the 4-D observed point p_{a_b} in the 4-D world-coordinate system $x_w y_w z_w w_w$. As the principal vanishing point vp_{w_b} moves to a new principal vanishing point vp_w, the other three vanishing points vp_x, vp_y and vp_z can be estimated using the position of the operated principal vanishing point vp_w as follows:

$$vp_x = \left(x_{vp_x}, y_{vp_x}, z_{vp_x} \right)$$

$$= \left(-\frac{1}{x_{vp_w}} \left\{ \left(\frac{h}{k}\right)^2 + y_{vp_w}{}^2 \right\}, y_{vp_w}, 0 \right),$$

$$vp_y = \left(x_{vp_y}, y_{vp_y}, z_{vp_y} \right)$$

$$= \left(0, -\frac{1}{y_{vp_w}} \left(\frac{h}{k}\right)^2, 0 \right), \tag{3}$$

$$vp_z = \left(x_{vp_z}, y_{vp_z}, z_{vp_z} \right)$$

$$= \left(x_{vp_w}, y_{vp_w}, -\frac{1}{z_{vp_w}} \left\{ \left(\frac{h}{k}\right)^2 + x_{vp_w}{}^2 + y_{vp_w}{}^2 \right\} \right).$$

Even if another principal vanishing point is moved, or if all the principal vanishing points are not displayed in 3-D space, the positions of the principal vanishing points can be estimated in the same manner as mentioned above. A movable region of the principal vanishing points vp_x and vp_y is restricted on the $x_s y_s$-plane and on the y_s-axis, respectively, while the principal vanishing points vp_z and vp_w can freely move in 3-D space. These restrictions mean that a 4-D upper direction, which corresponds to the y_e-direction, is kept in an upward vertical direction in the 4-D world-coordinate system.

Next, let us discuss the second processing step. We achieve the 4-D viewing direction control. That is, we consider movement of the 4-D observed point p_a along a 4-D spherical surface centered on the 4-D eye-point p_f (see Fig. 3). Parameters α, β and γ regarding the 4-D viewing direction are derived from Eq. (2) as follows:

$$\alpha = \tan^{-1} \frac{z_{vp_w}}{\sqrt{-z_{vp_z} z_{vp_w}}},$$

$$\beta = \tan^{-1} \frac{-x_{vp_w}}{\sqrt{-x_{vp_z} x_{vp_w}}}, \tag{4}$$

$$\gamma = \tan^{-1} \frac{-y_{vp_w}}{\sqrt{-y_{vp_z} y_{vp_w}}}.$$

Substituting the coordinate values of the principal vanishing points vp_x, vp_y, vp_z and vp_w estimated from the above-mentioned first processing step, the coordinate values of the Eq. (3) as an example, to Eq. (4), the corresponding parameters α, β and γ regarding the 4-D viewing direction are determined. The 4-D observed point p_{a_b} is represented as the transformed coordinates $(0, 0, 0, -r)$ in the 4-D eye-coordinate system $x_e y_e z_e w_e$, which is defined with the origin at the 4-D eye-point p_f and the w_e-axis in the direction from the 4-D observed point p_a to the 4-D eye-point p_f as described in Sect. 3. Therefore, the 4-D observed point p_a in the 4-D world-coordinate system $x_w y_w z_w w_w$ can be computed as the following equation:

$$
\begin{aligned}
p_a &= [x_{p_a} \quad y_{p_a} \quad z_{p_a} \quad w_{p_a} \quad 1] \\
&= [0 \quad 0 \quad 0 \quad -r \quad 1] T_{xz}^{-1}(\gamma) T_{yz}^{-1}(\beta) T_{xy}^{-1}(\alpha) T_t^{-1}(-p_f),
\end{aligned}
\tag{5}
$$

where the transformation matrix T_t represents the 4-D translation matrix, and the transformation matrices T_{xz}, T_{yz} and T_{xy} represent the 4-D rotation matrices around xz-, yz- and xy-planes, respectively. When the 4-D viewing direction is beyond the range of $-\pi/2 \leq \alpha, \beta, \gamma \leq +\pi/2$, the appropriate 4-D viewing direction is estimated by comparing the before and the after its change.

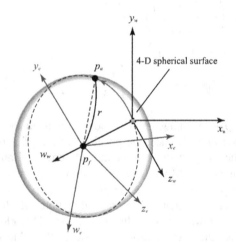

Fig. 3. Movement of the 4-D observed point along a 4-D spherical surface.

With the 4-D interaction method mentioned above, we can smoothly change the 4-D viewing direction at an arbitrary 4-D eye-point. The final 4-D visualization is achieved by introducing the 4-D observed point p_a of Eq. (5) into Eq. (1) shown in Sect. 3.

5   Construction of Interactive System for 4-D Space Exploration

In this section, we describe our interactive system that enables a user to intuitively explore 4-D space. As shown in Fig. 4, the developed system consists of commercially available products such as a personal computer (HP, Intel Core i7 3.90 GHz, 8 GB

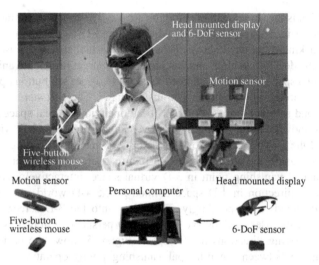

Fig. 4. Configuration of an interactive system.

RAM, NVIDIA GeForce GTX 680) installed Windows 8 (Microsoft), a motion sensor (ASUS Xtion Pro Live), a head mounted display with a built in a 6-DoF sensor (Vuzix Wrap 1200VR) and a five-button wireless mouse (ELECOM). Our proposed algorithm was implemented in C#, OpenTK, OpenNI and OpenCV under Visual Studio (Microsoft). The system guarantees real-time performance (60 frames per second) with interactive visualization of the 4-D scene including 45 or less hypercubes.

3-D virtual space seen from the head mounted display as a side by side stereoscopic image coincides with the $x_s y_s z_s$-space in the 3-D screen-coordinate system $x_s y_s z_s w_s$. Moreover, 3-D virtual space and real space have the same scale. In this work, we set the 3-D screen which has a dimension of 300 mm × 300 mm × 300 mm in 3-D virtual space. The 3-D perspective drawings of 4-D data and the principal vanishing points are displayed onto the 3-D screen and in 3-D virtual space on the head mounted display, respectively. In order to observe these to freely walk in 3-D virtual space, we associated a location and a viewing direction in 3-D virtual space with the user's head position and orientation in real space, which are measured using the motion sensor and the 6-DoF sensor built into the head mounted display, respectively. The user's head tracking data of the motion sensor is defined in the right-hand IR camera-coordinate system $x_c y_c z_c$, the origin of which lies at the IR camera of the motion sensor, and where the z_c-axis is in the direction from the IR camera to the user. We convert them to data of 3-D virtual space. By this transformation, the user's head position in real space corresponds to the location in 3-D virtual space. Therefore, the user can observe the 3-D perspective drawing of 4-D data and the principal vanishing points from various positions and directions in 3-D virtual space.

A 3-D cursor displayed in 3-D virtual space on the head mounted display has a function to indicate an arbitrary principal vanishing point in 3-D virtual space. We associated the 3-D cursor motion in 3-D virtual space with the user's hand motion in real space, which is measured using the motion sensor. The user's hand tracking data of the motion sensor in real space is converted to data in 3-D virtual space as well as the user's head tracking mentioned above. Thus, in 3-D virtual space, the user moves the 3-D cursor to the desired position, and can indicate the target principal vanishing point. Moreover, in real space, the user is allowed to take a two-step left button operation with the five-button wireless mouse holding in his/her hand. The first step is a click operation. The second step is a drag operation. These operations in real space are used for selection and movement of one principal vanishing point in 3-D virtual space, respectively. If the moving 3-D cursor and the target principal vanishing point overlap with each other, using the above-mentioned mouse operation, the user can pick and move the principal vanishing point in 3-D virtual space, interactively. As a result of this, the viewing direction in 4-D space defined in the 4-D world-coordinate system $x_w y_w z_w w_w$ is changed, correspondingly. Therefore, onto the 3-D screen on the head mounted display, the user can observe various 3-D perspective drawings of 4-D data from different viewing directions in 4-D space. Figure 5 shows the basic correspondence relationship between the principal vanishing points operation in 3-D virtual space and the viewing direction change in 4-D space. We put a hypercube at the position $(0, 0, 0, -1.5)$ on w_w-axis of the 4-D world-coordinate system and visualized the hypercube from the 4-D eye-point which lies at the origin of the 4-D world-coordinate system. We gave eight different colors to each cell of the hypercube. In order to clearly visualize edges inside the 3-D perspective drawing, we rendered semi-transparent faces with a reticular stipple pattern. Firstly, when the 4-D observed point is at the position $(0, 0, 0, -1.5)$ on w_w-axis, the principal vanishing point vp_w and the 3-D perspective drawing of the hypercube are displayed at the origin of the $x_s y_s z_s$-space corresponding to 3-D virtual space and on the 3-D screen, respectively. When the user moves the principal vanishing point vp_w in the directions of principal coordinate axes of 3-D virtual space which coincide with x_s-, y_s- and z_s-directions of the 3-D screen-coordinate system, the 4-D viewing direction changes toward to x_w-, y_w- and z_w-directions of the 4-D world-coordinate system, respectively. In accordance with each operation, the hypercube gradually moves outside of the 4-D viewing field. For example, the viewing direction in 3-D space is roughly composed of a combination of horizontal and vertical directions, while the viewing direction in 4-D space includes another direction. So, especially, the operation of the principal vanishing points vp_z and vp_w in the z_s-direction will be a key to master the 4-D viewing direction control. Therefore, from the correspondence between the 4-D viewing direction and the principal vanishing points operation, it is expected that the user will be able to enhance spatial understandings of 4-D space.

In addition to the 4-D viewing direction control, the system is allowed to simultaneously move both the eye-point and the observed point back and forth along the viewing direction in 4-D space. The forward and backward movements in 4-D space are associated with each click operation of two side buttons with the five-button wireless mouse in real space, respectively. Therefore, the user can freely fly through 4-D space, while changing the moving direction in 4-D space via the principal vanishing points operation in 3-D virtual space.

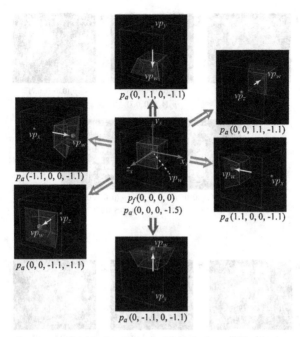

Fig. 5. Correspondence relationship between the principal vanishing points operation in 3-D virtual space and the viewing direction change in 4-D space. The parameters k, h and f regarding the 4-D viewing field are given as 0.5, 0.5, and 100, respectively.

6   Visualization of 4-D Solid Scene Using 4-D Space Exploration System

In this section, we demonstrate some 4-D scenes such as insides of 4-D solids, multiple 4-D solids and a 4-D maze using the developed system, and show that the system can provide novel experiences and understandings of 4-D space different from our previous system [12].

6.1   Inside of 4-D Solids

By putting the 4-D eye-point inside a 4-D solid and changing the 4-D viewing direction in the 4-D solid, we can visualize various insides of the 4-D solid clipped by the 3-D screen. Figures 6 and 7 show the insides of a 24-cell and a 120-cell visualized using this manner. We can understand that the 24-cell and the 120-cell is constructed by octahedrons and dodecahedrons, respectively, and that each surface of the 4-D solid is shared by two different cells, generally. In our system, eight different colors are given to each cell of the 4-D solid. This visualization will help us to perceive geometric characteristics and structure of the 4-D solid.

In addition, although the 4-D viewing direction is different in 4-D space, the obtained principal vanishing points are the same in pictures 2 and 3 of Figs. 6(a) and 7(a),

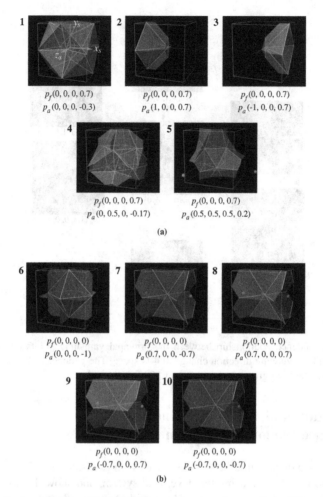

Fig. 6. Visualization of a 24-cell by 4-D viewing direction changes from 4-D eye-points inside the 24-cell. The 24-cell is placed at the origin of the 4-D world-coordinate system. The parameters k, h and f regarding the 4-D viewing field are given as 0.5, 0.5 and 100, respectively. (a) The 4-D eye-point is at the position $(0, 0, 0, 0.7)$ in the 4-D world-coordinate system. (b) The 4-D eye-point is at the origin of the 4-D world-coordinate system.

and pictures 7 and 9, and 8 and 10 of Figs. 6(b) and 7(b). This happens when the user turns the 4-D viewing direction 180 degrees in 4-D space. Understanding of these relationships will lead to intuitively handle turning about a motion in 4-D space.

6.2   Multiple 4-D Solids

While most previous research has mainly handled only one type of 4-D solid with a fixed eye-point and viewing direction in 4-D space, our system can provide smooth interaction with the 4-D scene including multiple 4-D solids where users can freely

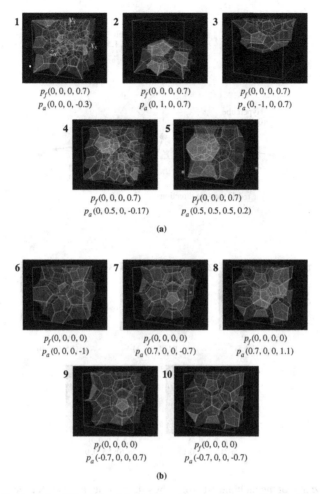

Fig. 7. Visualization of a 120-cell by 4-D viewing direction changes from 4-D eye-points inside the 120-cell. The 120-cell is placed at the origin of the 4-D world-coordinate system. The parameters k, h and f regarding the 4-D viewing field are given as 0.5, 0.5 and 100, respectively. (a) The 4-D eye-point is at the position (0, 0, 0, 0.7) in the 4-D world-coordinate system. (b) The 4-D eye-point is at the origin of the 4-D world-coordinate system.

change the eye-point and viewing direction. For example, as shown in Fig. 8(a), there are a hypercube, a 5-cell and a 16-cell at the positions (0, 0, 0, 1.5), (1.5, 0, 0, 0) and (0, 1.5, 0, 0) in the 4-D world-coordinate system, respectively, so that they surround the 4-D eye-point at the origin of the 4-D world-coordinate system. Figure 8(b) depicts an image sequence of 3-D perspective drawings obtained by turning the 4-D viewing direction to each 4-D solid. Initially, the 3-D perspective drawing of the hypercube is shown at the center of the 3-D screen. By moving principal vanishing points vp_w, vp_x and vp_y, respectively, the 4-D viewing direction changes (see Fig. 8(a)) and the 5-cell and the 16-cell come into the 3-D screen in sequence (see Fig. 8(b)).

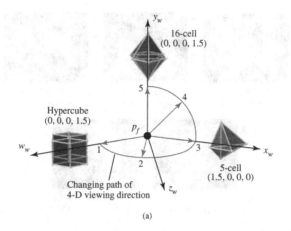

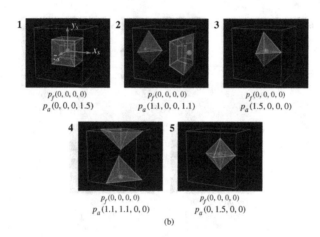

Fig. 8. Visualization of the multiple 4-D solids by 4-D viewing direction changes. (a) Layout of the 4-D solids and the changing path of the 4-D viewing direction. (b) Image sequence obtained by 4-D viewing direction changes. The parameters k, h and f regarding the 4-D viewing field are given as 0.5, 0.5 and 100, respectively. The numbers of the pictures correspond to the numbers of the 4-D viewing direction in (a).

Moreover, we demonstrate the visualization of the multiple 4-D solids with 4-D fly-through actions (see Fig. 9). As shown in Fig. 9(a), we use the 4-D scene which includes a hypercube at the origin of the 4-D world-coordinate system and a 16-cell at the position $(0, 0, 0, 3.5)$ in the 4-D world-coordinate system (see Fig. 9(a)). Figure 9(b) depicts an image sequence of 3-D perspective drawings of 4-D solids obtained by traveling around 4-D space. In the initial projection obtained from the 4-D eye-point $p_f(0, 0, 0, 1.5)$ on the w_w-axis, the 3-D perspective drawings of the hypercube and the 16-cell have an overlap with each other in the 3-D screen as shown in picture 1 of Fig. 9(b). This means that the 16-cell is occluded by the hypercube in 4-D space. Then, by changing the 4-D viewing direction and moving the 4-D eye-point from on the w_w-axis toward the 16-cell

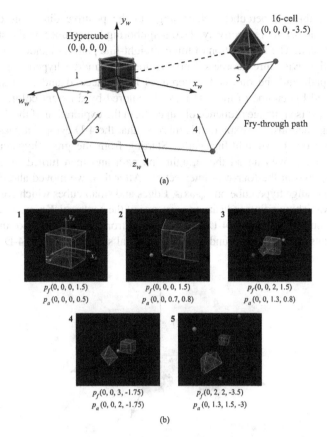

Fig. 9. Visualization of the multiple 4-D solids by 4-D fly-through actions. (a) Layout of the 4-D solids and the fly-through path. (b) Image sequence obtained by 4-D fly-through actions. The parameters k, h and f regarding the 4-D viewing field are given as 0.5, 0.5 and 100, respectively. The numbers of the pictures correspond to the numbers of the 4-D viewing direction in (a).

(see Fig. 9(a)), these 3-D perspective drawings are separated off on the 3-D screen, and align in the direction of the principal vanishing point vp_w (see Fig. 9(a)). Therefore, we can find that the hypercube and the 16-cell locate on different positions on the w_w-axis of the 4-D world-coordinate system. In this way, the 4-D fly-through action will be useful to understand the layout of the 4-D solids in the 4-D scene, even if it is intricate with occlusions.

6.3   4-D Maze

We show a 4-D maze exploration using 4-D fly-through actions. As shown in Fig. 10 (a), a 4-D maze is formed so that three straight-line paths on x_w-, y_w- and z_w-axes in the 4-D world-coordinate system intersect at the origin of the 4-D world-coordinate system. The first path is constructed by connecting five hypercubes, which align in the negative and positive directions on x_w-axis. The second and third paths are constructed by connecting three hypercubes, which align in the positive directions on z_w- and w_w-axes from the origin, respectively. Two neighboring hypercubes on the straight-line path share one cell. The intersection of three straight-line paths coincides with the four-way intersection with three corners, and includes the middle hypercube in the first straight-line path and the one end hypercubes in the second and third straight-line paths. Each end hypercube of the 4-D maze is colored by different color.

Figure 10 (b) is an image sequence obtained from the exploration of the 4-D maze by 4-D fly-through actions. We explore this maze so that the 4-D eye-point passes through the hypercubes on the straight-line paths. Starting from the green hypercube on the w_w-axis, we first moved ahead the straight-line path and then turned to the positive direction of x_w-axis at the four-way intersection. After that, we moved ahead again and stopped at the orange hypercube on x_w-axis. Edges and small cubes which concentrate to the principal vanishing points indicate an existence of the next way. Knowledge regarding orthogonality and parallelism in 4-D space obtained from characteristics of the principal vanishing points will help us to understand the spatial structure of such 4-D maze.

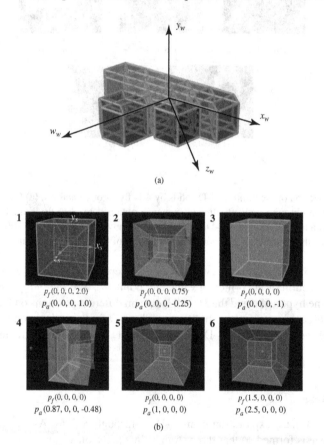

Fig. 10. Exploration of the 4-D maze by 4-D fly-through actions. (a) Appearance of the 4-D maze visualized using the parallel projection. (b) Image sequence obtained from the exploration of the 4-D maze by 4-D fly-through actions. The parameters k, h and f regarding the 4-D viewing field are given as 0.5, 0.5 and 2.5, respectively.

7 User Experiments with Interactive 4-D Space Exploration System

In this section, we describe a user experiment in order to examine whether the proposed system can provide sufficient usability and learnability to handle the 4-D viewing direction in 4-D space. In the experiment, subjects spot the target 4-D object in 4-D space, while changing the 4-D viewing direction via the principal vanishing points operation. If the subjects can reduce response time to complete the tasks and show a positive impression for the interaction, it will be proof of the usability and learnability of the proposed system.

7.1 Method

Participants. Eleven male subjects and one female subject participated in the experiment without monetary compensation. The subjects were students or graduate students who majored in science and engineering at Waseda University. Their mean age was 21.4 years. Seven male subjects had some previous knowledge of 4-D space and objects, since they had observed the hypercube with our previous interactive system [12]. However, they had never experienced the 4-D viewing direction control and had no knowledge of the current experiment. The other four male and one female subjects had never had any 4-D interaction. They did not have any knowledge of 4-D space and the proposed interactive system. In addition, they had no experience of an immersive virtual reality system using the head-mounted display, either.

We divided the twelve subjects into two groups based on their previous knowledge. Thus, group A had seven male experienced subjects of mean age 21.6 years. Group B had five beginner subjects of mean age 21.0 years. Both groups were given the same tasks.

Apparatus. We ran the experiment inside a meeting space of our university building (see Fig. 4). Subjects used the head-mounted display and the five-button wireless mouse to negotiate the experimental tasks. The 3-D screen was placed 1.8 meters away from the motion sensor. This configuration provided the enough spatial resolution to track the subjects' head and hand positions. The 4-D scene was rendered in 60 frames per second, consistently.

Stimuli and Tasks. The experiment was performed in two stages. In each stage, firstly, subjects engaged in 20 trials, then following a three-minute break repeated the same 20 trials. Thus, the subjects performed 80 trials in total.

Stimuli were simple 4-D scenes. We constructed 20 scenes for each stage. We used the scenes twice in each stage. Each 4-D scene included one hypercube as the target object in 4-D space. The hypercube had a dimension of $1 \times 1 \times 1 \times 1$ and we gave eight different colors to each cell as well as Fig. 5. The position of the target hypercube was different in each scene. In the first stage, the target was placed at distance 2 from the origin of the 4-D world-coordinate system so that at least one vertex of the target hypercube was included in the 4-D viewing field when the subjects started the trial.

In order to make 20 scenes, we randomly chose 20 positions in 4-D space. The mean angle between the initial 4-D viewing direction which coincided with the negative direction of the w_w-axis and the direction to the target hypercube was 42.4 degrees. In the second stage, the target hypercube was placed at distance 2 from the origin of the 4-D world-coordinate system as well as the first stage. However, the position of the target hypercube was behind the 4-D eye-point when the subjects started the trials. We chose 20 positions in 4-D space to make 20 scenes, again. In the second stage, the mean angle between the initial 4-D viewing direction and the direction to the target hypercube was 116.8 degrees.

Tasks of the experiment were to spot the target hypercube in 4-D space. In each trial, the subjects were asked to change the 4-D viewing direction from the negative direction of the w_w-axis to the center of the target hypercube in 4-D space. The 4-D eye-point was fixed at the origin of the 4-D world-coordinate system. In other words, the goal of the trial was to capture the target hypercube at the center of the 3-D screen. In each trial, the subjects could check whether or not they completed the task at any time they desired. When the subjects checked their answer, the system calculated an anguler error which corresponds to the angle between the current 4-D viewing direction and the direction to the target. If the angular error was under a threshold, the subjects proceeded to the next trial. Otherwise, the subjects had to continue the current trial until they reduced the angular error sufficiently. In the present experiment, we determined the threshold of 15 degrees based on results of our preliminary experiment.

Procedure. Before the subjects started the trials, we gave each subject a brief expla-nation on 4-D space and objects, the visualization model and the interactive system. This explanation took approximately 10 to 15 minutes. Firstly, we defined 4-D space and the structure of the hypercube; 4-D space was defined by four principal coordinate axes which were orthogonal to each other; the hypercube of the 4-D solid was con-structed by eight cells (3-D cubes) and each surface is shared by two different cells. Then, we explained the 4-D visualization model based on the analogy with 3-D space; the hypercube in 4-D space is displayed as the 3-D perspective drawing into the 3-D screen by the projection method; the principal coordinate axes in 4-D space were displayed as the principal vanishing points in 3-D space by the projection method. After that, we showed how to use the interactive system by displaying the 3-D per-spective drawing, principal vanishing points and 3-D cursor in 3-D virtual space on the head-mounted display. The 3-D cursor was used to pick and move the specific principal vanishing point. The subjects could move around 3-D virtual space by walking around real space.

After the explanation was finished, we gave the subjects five minutes to practice using the 4-D viewing direction control using the interactive system. We used the 4-D scene where the target hypercube was placed at the position $(0, 0, 0, -2)$ in the 4-D world-coordinate system as an example. The 4-D eye-point and the initial 4-D viewing direction were as same as the condition of the task. The subjects freely handled the 4-D viewing direction by the principal vanishing points operation and saw the hypercube being in and out of the 4-D viewing field. Note, we adjusted the stereoscopic parallax to display the 3-D image on the head-mounted display in this practice phase.

The subjects received a three minute break after the practice run. They, then, proceeded to the first stage of the trials, followed three-minute break, after which second stage of the trials was conducted. These two stages took approximately 45 to 60 minutes. Although every subject challenged the same trials, the order of the trials was shuffled to reduce an order effect. We recorded their response time, accuracy (the final angular error) and action histories for the 4-D viewing direction control in each trial. These results were used to evaluate the usability and the learnability of the proposed system.

After the subjects finished tasks, they answered a questionnaire to elucidate their impression of the proposed system. The questionnaire included the following seven questions and a free comment column.

1. How was overall impression of the system, positive or negative?
2. How easy to learn operations of the system?
3. How easy to use the system?
4. How easy to turn the 4-D viewing direction to the target hypercube according to your plans?
5. Did you understand the relationship between operations of the principal vanishing points and the resulting changes on the 3-D screen?
6. Did you enhance your sense of direction in 4-D space?
7. Did you reduce mistakes of the 4-D viewing direction control during the trials?

The rating used a seven-point scale. The subjects answered a score between "minus three" and "plus three" for each question, a score of "plus three" being at the positive end of the spectrum while the score of "zero" indicates neutral impression. A score of "minus three" is the most negative response. These subjective reports were used to assess the usability of the proposed system.

7.2   Results

960 trials were run in total by twelve subjects. We regarded 945 trials as valid results, because, in 15 trials, the motion sensor suddenly failed to properly track the subjects' head and hand positions.

Response Time. Firstly, we describe the results of 40 trials of the first stage. Figure 11 shows the mean response time of the trials. In order to remove noise of time data, we plotted moving average of each four trials with standard errors displayed as error bars. Initially, the mean response time for group A was 64.1 seconds. The subjects reduced their response time in the first five trials. Then, their performance stabilized at the mean response time of 12.3 seconds. Group B yielded similar results. Their initial mean response time was 115.8 seconds. Their response time rapidly dropped in the first five trials. This drop gradually continued in the sixth to the 15th trials. Then, their performance stabilized at the mean response time of 15.1 seconds.

Improvement in subjects of group B was also confirmed in their action histories. Figure 12 shows time courses of the angular error between the 4-D viewing direction and the direction to the target hypercube. Because subjects of group B followed similar

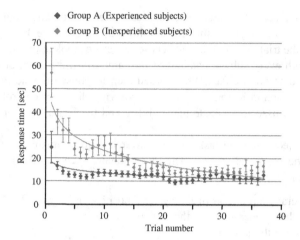

Fig. 11. Relationship between the mean response time and the trial number in the first stage.

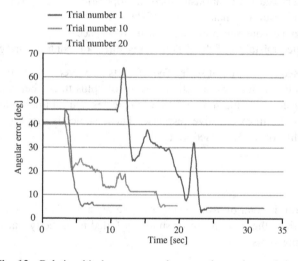

Fig. 12. Relationship between angular error decreasing and time.

trends, we plotted the first, tenth and 20th trials of one subject as representative results. Increases in angular error represent subjects turning the 4-D viewing direction to the wrong direction in 4-D space. The results showed that as the subject practiced, they reduced the number of incorrect operations and time spent to correct the misguiding.

Next, we describe the results of 40 trials of the second stage. Figure 13 shows the mean response time of the trials. Subjects gave similar results to the first stage. Both groups reduced their response time in the first ten trials, after which their performance stabilized. The mean response times of the 11th to the 40th trials were 23.2 seconds for group A and 33.7 seconds for group B. The subjects succeeded in searching for and spotting the target hypercube behind the 4-D eye-point in 4-D space.

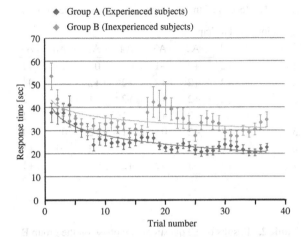

Fig. 13. Relationship between the mean response time and the trial number in the second stage.

Accuracy. We also describe the subjects' accuracy of the 4-D viewing direction control by calculating the mean angular error of 80 trials for each group. The mean angular error of group A was 5.37 degrees. The mean angular error of group B was 6.54 degrees. These angular errors correspond the target hypercube displacement from the center of the 3-D screen.

Subjective Impression. Finally, we show the subjects' impression of the 4-D inter-action. Tables 1 and 2 summarize full data of the results of the questionnaire for groups A and B. The boxplot for full data shown in Tables 1 and 2 appears in Fig. 14. Question number 1 asked the overall impression. The median responses were a score of "plus two" for group A and a score of "plus three" for group B, respectively. Every subject had positive impressions of the proposed 4-D interaction. Question numbers 2 and 3 enquired as to the learnability and the usability of the interaction. The median response of group A were "plus two" for each question, respectively. The median response of group B were "plus one" and "plus three", respectively. No one gave negative responses. Question numbers 4 to 7 asked the impression of the 4-D viewing direction control. Question number 4 asked how easy the subjects could change the 4-D viewing direction to the hypercube. The mean responses were the score of "plus one" for both groups. Only subject B3 responded negatively. Question number 5 asked whether subjects can grasp the relationship between their operation and the resulting 4-D viewing direction changes. The median responses were "plus one" for group A and "plus two" for group B, respectively. Question number 6 asked whether they improved their perception of 4-D space. The number of the neutral and negative responses were high compared to the other six questions in this question. We discuss this point in the next section. Question number 7 asked whether they felt that they could improve their performance by themselves. Both groups showed the median score of the "plus two" and no one responded negatively. Summarily, these results suggest that subjects found the interactive system easy to learn and use.

Table 1. Results of the questionnaire phase on the group A.

Question No.	Participant						
	A1	A2	A3	A4	A5	A6	A7
1	+2	+2	+1	0	0	+2	+1
2	+2	+2	+2	+2	+2	+2	0
3	+2	+2	+1	+2	+1	+1	+2
4	+3	+2	+1	+1	+3	+1	+1
5	+2	+2	0	+1	+2	0	−1
6	+1	+2	0	−1	+2	−1	+1
7	+2	+3	+2	+2	0	+2	+1

Table 2. Results of the questionnaire phase on the group B.

Question No.	Participant				
	B1	B2	B3	B4	B5
1	+3	+3	+3	+3	+3
2	+2	+3	0	0	+1
3	+1	+3	+2	+3	+3
4	0	+1	−3	+1	+1
5	+2	+2	+1	−2	+2
6	0	−1	0	0	+2
7	+2	+3	+3	+1	+2

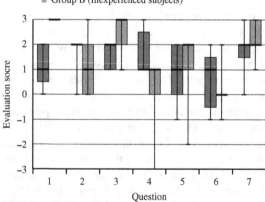

Fig. 14. Results of the questionnaire about usability and learnability of the developed system.

7.3 Discussions

Usability and Learnability. Both groups showed a similar pattern of results in the first stage. Although the learning speeds were different for each of the subjects, they reduced their response time in early stage of the first 20 trials and consistently performed smooth 4-D viewing direction control in the latter 20 trials. The interpretation of this pattern of performance is that subjects rapidly learn the correspondences between the principal vanishing points operation and the 4-D viewing direction changes in the first five to ten trials. Then, they use that obtained understanding over the trials. This interpretation is confirmed by their positive impression shown in the questionnaire scores and comments. Subjects found the task difficult at the beginning of the experiment. However, once they understand the correspondence between the principal vanishing point operation and its result visualized on the 3-D screen, trials become easier to complete.

In the second stage, the subjects initially searched the target by looking around 4-D space. Once they found the target hypercube in the 4-D viewing field, they captured it into the center of the 3-D screen by handling the 4-D viewing direction as well as the first stage. We find that subjects' response times are roughly twice as long as the first stage. Because the target hypercube is located behind the 4-D eye-point, the subjects have to turn back the 4-D viewing direction in 4-D space to find the hypercube. In order to achieve this action, the subjects need to operate at least two principal vanishing points. Moreover, their amount of interaction is approximately twice as much as the first stage. Considering these conditions, the increase in response time is not unexpected.

In summary, we conclude that the proposed system makes it easy to learn and handle the 4-D viewing direction control, even if the user does not have previous knowledge of 4-D space and objects. The system is suitable for looking around and finding 4-D objects in 4-D space.

Evaluation of Accuracy. Based on the results, there were approximately six degrees of the angular error, when subjects completed the task. This error can be explained by a combination of the following interpretations. Firstly, we consider the primary cause to be that it is difficult for the users to completely fix their hand in the air. This may be particularly pronounced when subjects slightly move the principal vanishing points to fine-adjust the 4-D viewing direction. In fact, many subjects commented on this point.

Secondly, there is approximately 150 milliseconds latency in the motion sensor due to the smoothing filter reducing the positional noise of the head and hand tracking. Although there has been no feedback regarding the latency, it is possible that this affects the interaction.

Thirdly, the error is due to the subjects confusing the depth of the center of the 3-D screen on the stereoscopic visualization, for example, when the opaque part of the 3-D perspective drawings of the target hypercube occludes the center of the 3-D screen which is indicated as a cross-point of the x_s-, y_s- and z_s-axes displayed in 3-D virtual space.

Effect of Previous Knowledge. In the both stages, the response time of group A was significantly shorter than group B ($p = 3.28 \times 10^{-3} < 0.05$ for the first stage, and $p = 1.37 \times 10^{-5} < 0.05$ for the second stage). We also confirm that the performance of group A was more stable among the trials than group B. In addition, the scores of Question number 4 of the questionnaire for group A were significantly higher than those of group B ($p = 0.0254 < 0.05$). These results suggest that it is easy for subjects of group A to solve the task, compared to group B. We speculate that this difference in performance occurs due to differences in subjects' previous knowledge of 4-D space and objects. Because subjects of group A have observed the hypercube by controlling the 4-D eye-point along the 4-D spherical surface via the principal vanishing points operation [12], they may naturally grasp the 4-D viewing direction control in the present experiment. This is proof, at least in part, that humans were able to improve their 4-D spatial perception with the various experiences in the 4-D interactive environment.

Limitation of Present Study. We confirm that the proposed system makes it easy to learn and provide the sufficient usability to handle the 4-D viewing direction, however, some feedback comments highlight limitations of the system. For example, one subject felt that the viewing field on the head-mounted display was narrow. Some subjects found the view bobbing in 3-D virtual space, which occurred especially when their hand occluded their head from the motion sensor, annoying. Combined with a discussion about the accuracy, the improvement of the tracking system and the stereoscopic display system will be the subject of further work, with the aim to provide more comfortable interaction.

As shown by Question number 6 of the questionnaire, in the present experiment, we cannot confirm whether the subjects obtain the general spatial perception of 4-D space. This is because we designed the experiment to primarily assess the usability of the proposed system. In order to evaluate the human 4-D spatial perception, it will be necessary to plan the cognitive test to examine whether subjects grasp the spatial structure of the 4-D objects and the 4-D scenes such as 4-D maze. The proposed system can be used as a test bed for such cognitive tests in the future work.

8 Conclusions

In this chapter, we described the novel 4-D interaction technique that enables the user to intuitively control the viewing direction in 4-D space via the principal vanishing points operation in 3-D space. Using an interactive interface, our system can visualise the 4-D scenes such as the inside of the 4-D solids, the multiple 4-D solids and the 4-D maze. In such 4-D scenes, the principal vanishing points are useful in understanding the orthogonality and the parallelism in 4-D space. In order to assess the usability of the proposed interactive system, the user experiment was conducted with simple 4-D scenes. We examined whether the subjects correctly handled the 4-D viewing direction to the predetermined direction. The results showed that the principal vanishing points operation provided the sufficient usability for the 4-D viewing direction control. Subjects easily learned the interaction and intuitively looked around 4-D space, even if they had no previous knowledge of 4-D space and objects.

General human spatial perception in 4-D space was outside the scope of this research, due to experiment design. Future work may include spatial cognitive tests in

4-D space such as mental rotation, spatial orientation and maze exploration. The proposed system will be used as the test bed of the cognitive tests. In addition, we will try to apply the system to modeling and editing of 4-D solids and data.

Acknowledgments. This work was supported in part by Global COE Program "Global Robot Academia" and Grants for Excellent Graduate Schools from the Ministry of Education, Culture, Sports, Science and Technology of Japan.

References

1. Abbott, E.A.: Flatland: A Romance of Many Dimensions. New American Library, New York (1984)
2. Aguilera, J.C.: Virtual reality and the unfolding of higher dimensions. In: Proceedings of SPIE. Stereoscopic Displays and Virtual Reality Systems XIII vol. 6055, pp. 605–612, January 2006
3. Banchoff, T.F.: Beyond the Third Dimension: Geometry, Computer Graphics, and Higher Dimensions (Scientific American Library Series). W. H. Freeman & Company, New York (1990)
4. Chu, A., Fu, C.W., Hanson, A.J., Heng, P.A.: GL4D: A GPU-based architecture for interactive 4D visualization. IEEE Trans. Visual. Comput. Graph. **15**(6), 1587–1594 (2009)
5. Davis, P.J., Hersh, R., Marchisotto, E.A.: Four dimensional intuition. In: The Mathematical Experience, Study Edition, pp. 442–447. Birkhäuser Boston (October 1995)
6. Dewdney, A.K.: Computer recreations: a program for rotating hypercubes induces four-dimensional dementia. Sci. Am. **254**, 8–13 (1986)
7. D'Zmura, M., Colantoni, P., Seyranian, G.D.: Virtual environments with four or more spatial dimensions. Presence **9**(6), 616–631 (2000)
8. Hanson, A.J.: Computer graphics beyond the third dimension. In: Course Notes for SIGGRAPH 1998, Course 46, July 1998
9. Hanson, A.J., Heng, P.A.: Illuminating the fourth dimension. IEEE Comput. Graph. Appl. **12**(4), 54–62 (1992)
10. Hausmann, B., Seidel, H.P.: Visualization of regular polytopes in three and four dimensions. Comput. Graph. Forum **13**(3), 305–316 (1994)
11. Hollasch, S.R.: Four-Space Visualization of 4D Objects. M.S. Dissertation, Arizona State University, August 1991
12. Miwa, T., Sakai, Y., Hashimoto, S.: Four-dimensional eye-point control by principal vanishing points operation and its evaluations. J. Soc. Art Sci. **12**(4), 162–174 (2013). (in Japanese)
13. Miwa, T., Sakai, Y., Hashimoto, S.: Four-dimensional viewing direction control by principal vanishing points operation and its application to four-dimensional fly-through experience. In: Proceedings of the 25th Australian Computer-Human Interaction Conference: Augmentation, Application, Innovation, Collaboration, pp. 95–104, November 2013
14. Miyazaki, K., Ishihara, K.: Four-Dimensional Graphics. Asakura-Shoten Publishers (September 1989), (in Japanese)
15. Neophytou, N., Mueller, K.: Space-time points: 4D splatting on efficient grids. In: Proceedings of the 2002 IEEE Symposium on Volume Visualization and Graphics. pp. 97–106, October 2002

16. Sakai, Y., Hashimoto, S.: Interactive four-dimensional space visualization using five-dimensional homogeneous processing for intuitive understanding. J. Inst. Image Inf. Telev. Eng. **60**(10), (108)1630–(125)1647 (2006)
17. Sakai, Y., Hashimoto, S.: Four-dimensional space-time visualization for understanding three-dimensional motion. J. Inst. Image Electron. Eng. Japan **36**(4), 371–381 (2007)
18. Sakai, Y., Hashimoto, S.: Four-dimensional space visualization with four-dimensional viewing field control. Bull. Soc. Sci. Form **21**(3), 274–284 (2007). (in Japanese)
19. Sakai, Y., Hashimoto, S.: Four-dimensional geometric element definitions and interferences via five-dimensional homogeneous processing. J. Visual. **14**(2), 129–139 (2011)
20. Sakai, Y., Hashimoto, S.: Four-dimensional mathematical data visualization via embodied four-dimensional space display system. Forma **26**(1), 11–18 (2011)
21. Wang, W.M., Yan, X.Q., Fu, C.W., Hanson, A.J., Heng, P.A.: Interactive exploration of 4D geometry with volumetric halos. In: The 21th Pacific Conference on Computer Graphics and Applications - Short Papers, pp. 1–6, October 2013
22. Woodring, J., Wang, C., Shen, H.W.: High dimensional direct rendering of time-varying volumetric data. In: Proceedings of the 14th IEEE Visualization 2003, pp. 417–424, October 2003
23. Yan, X., Fu, C.W., Hanson, A.J.: Multitouching the fourth dimension. IEEE Comput. **45**(9), 80–88 (2012)

Touch the 3rd Dimension! Understanding Stereoscopic 3D Touchscreen Interaction

Ashley Colley[1]([⊠]), Jonna Häkkilä[1], Johannes Schöning[2],
Florian Daiber[3], Frank Steinicke[4], and Antonio Krüger[3]

[1] CIE, University of Oulu, Oulu, Finland
{ashley.colley,jonna.hakkila}@cie.fi
[2] Hasselt University, tUL – iMinds, Diepenbeek, Belgium
johannes.schoening@uhasselt.be
[3] DFKI GmbH, Saarbrucken, Germany
{florian.daiber,krueger}@dfki.de
[4] Department of Informatics, University of Hamburg, Hamburg, Germany
steinicke@informatik.uni-hamburg.de

Abstract. The penetration of stereoscopic 3D (S3D) output devices is becoming widespread. S3D screens range in size from large cinema screens, to tabletop displays, TVs in living rooms, and even mobile devices are nowadays equipped with autostereoscopic 3D screens. As a consequence, the requirement for interacting with 3D content is also increasing, with "touch" being one of the dominant input methods. However, the design requirements and best practices for interaction with S3D touch screen user interfaces are only now evolving. To understand the challenges and limitations S3D technology brings to interaction design, and in particular the additional demands they place on the users of such interfaces, we present our research that considers stereoscopic touch screens of different sizes in static and mobile usage contexts. We report on perceptual, cognitive and behavioral aspects revealed by user studies and present interaction design requirements based on the findings.

Keywords: Stereoscopy · 3D user interfaces · Tabletops · Mobile user interfaces · User studies

1 Introduction

Display and visualization technologies are advancing on various frontiers, and today we see both increasingly bigger and smaller screens with ever improving resolutions. One emerging trend is the use of stereoscopic 3D (S3D). Stereoscopic 3D is probably most familiar to large audiences from 3D movies, and 3DTVs and S3D mobile devices are already mass-market products. Autostereoscopic 3D displays which require no special glasses to experience the 3D effect can be found in such products as 3D cameras, mobile phones, tablets and game consoles such as the Nintendo 3DS. The S3D technologies create the illusion of depth, such that graphical elements are perceived to pop up from or sink below the surface level of the screen. With negative disparity (or parallax), the User Interface (UI) elements appear to float in front of the

© Springer International Publishing Switzerland 2015
T. Wyeld et al. (Eds.): OzCHI 2013, LNCS 8433, pp. 47–67, 2015.
DOI: 10.1007/978-3-319-16940-8_3

Negative disparity

Screen level

Positive disparity

Fig. 1. An autostereoscopic 3D display, where the content is displayed with negative (bird), zero (monkey) and positive (background image) disparity/parallax.

screen, and with positive disparity (or parallax), the UI elements appears behind the screen level (see Fig. 1).

So far, much of the research in the field of human-computer interaction (HCI) on S3D has focused on visual ergonomics and visual comfort (see e.g. [28, 29, 39]), and until now, research on interaction and user experience design for S3D has received less attention. However, recent research has emerged in various application domains for S3D, ranging from interactive experiences in S3D cinemas [16], to mobile games [11], as well as in investigating the interaction design [6, 9, 47, 48].

The 3rd dimension provides new degrees of freedom for designers, and the illusion of depth can be utilized both for information visualization as well as for interactive user interface (UI) elements. However, as the physical displays are still two-dimensional, it remains challenging to design interactive systems utilizing S3D. Especially, touch screen interaction is problematic with objects that appear visually in 3D but are unable to be touched. This mismatch between the visual and tangible perception can potentially increase the user's cognitive load, and make the systems slower to use and less easy to comprehend. Thus, it is important to investigate the effects that stereoscopic 3D can have on interaction compared to conventional 2D UIs. In order to successfully introduce novel designs for S3D UIs, we need to investigate the implications of S3D on the user interaction e.g. in terms of touch accuracy and resolution.

In this chapter, we investigate touchscreen-based interaction with S3D visualizations, and compare the differences between S3D and 2D interaction. By examining the utility of S3D touchscreen UIs for practical usage when the user is interacting with different size S3D displays, we provide recommendations to researchers, developers and designers of such UIs. Especially, we:

1. Provide a systematic analysis of the differences in interaction between 2D and S3D touch screens.
2. Investigate touch screen based S3D interaction in the context of both large and small screens, namely tabletops and mobile devices.
3. Report on differences between S3D touch screen interaction while on the move (walking) and when static (standing).
4. Provide design recommendations for designing interactive touch S3D systems.

2   Related Work

In this section we first briefly summarize the more general research on S3D in the area of HCI and then continue with a review of related work focusing on touch screen interaction with S3D.

2.1   HCI Research in S3D

Visual Experience. With research on mobile S3D UIs, much of the emphasis so far has been focused on the output visualizations – video, images and visual cues - rather than on interacting with the device. Jumisko-Pyykkö et al. [24] studied the users' experience with mobile S3D videos, and discovered that the quality of experience was constructed from the following: visual quality, viewing experience, content, and quality of other modalities and their interactions. Pölönen et al. [39] report that with mobile S3D the perceived depth is less sensitive to changes in the ambient illumination level than perceived naturalness and overall image quality. Mizobuchi et al. [34] report that S3D text legibility was better when text was presented at zero disparity on a positive disparity background image when compared to presenting it hovering above a background image that was at zero disparity. It has also been pointed out that scaling the S3D content to different size displays is not straightforward, and has an effect on perceptual qualities [2].

Cognition and Perception. Considering the effect of S3D on users' cognition, Häkkinen et al. [19] have investigated people's focal attention with S3D movies using eye tracking. When comparing the S3D and 2D versions of the movies, they found that whereas in 2D movie viewers tend to focus quickly on the actors, in the S3D version the eye movements were more widely distributed among other targets. In their FrameBox and MirrorBox, Broy et al. [5] identify that the number of depth layers has a significant effect on task completion time and recognition rate in a search task.

The role of visual cues has also been investigated, with some conflicting conclusions on their relative importance in depth perception. Mikkola et al. [33] concluded that the stereoscopic depth cues outperform the monocular ones in efficiency and accuracy, whilst Huhtala et al. [23] reported that for a find-and-select task in a S3D mobile photo gallery, both the performance and subjective satisfaction were better when the stereoscopic effect was combined with another visual effect, i.e. dimming. Kerber et al. [26] investigated depth perception in a handheld stereoscopic augmented reality scenario. Their studies revealed that stereoscopy has a negligible if any effect on a small screen, even in favorable viewing conditions. Instead, the traditional depth cues, in particular object size, drive depth discrimination. The perception of stereoscopic depth on small screens has also been examined by Colley et al. [9], who compared users' ability to distinguish stereoscopic depth against their ability to distinguish 2D size. More recently, Mauderer et al. [32] investigated gaze-contingent depth of field as a method to produce realistic 3D images, and analyzed how effectively people can use it to perceive depth.

Non-touchscreen Interaction Techniques. Besides touchscreen-based interaction, several other interaction methods have been applied to S3D interfaces. Teather et al. [45]

demonstrated a "fish tank" virtual reality system for evaluating 3D selection techniques. Motivated by the successful application of Fitts' law to 2D pointing evaluation, the system provided a testbed for consistent evaluation of 3D point-selection techniques. The primary design consideration of the system was to enable direct and fair comparison between 2D and 3D pointing techniques. To this end, the system presents a 3D version of the ISO 9241-9 pointing task. Considering a layered stereoscopic UI, Posti et al. [38] studied the use of hand gestures to move objects between depth layers.

Gestures with mobile phones have also been utilized as an interaction technique. In the context of an S3D cinema Häkkilä et al. [16] utilized mobile phone gestures as one method to capture 3D objects from the film. Using a large-scale stereoscopic display Daiber et al. [10] investigated remote interaction with 3D content on pervasive displays, concluding that physical travel-based techniques outperformed the virtual techniques.

Design Implications. The question of how the stereoscopic depth effect could be used in UI design has been investigated in several studies. Using S3D is perceived as a potential method of grouping similar content items or highlighting contextual information in a user interface [49]. There have also been design proposals where object depth within S3D UIs has been considered as an informative parameter e.g. to represent the time of the last phone call [18], and to identify a shared content layer in photo sharing [17]. Daiber et al. [11] investigated sensor-based interaction with stereoscopic displayed 3D data on mobile devices and presented a mobile 3D game that makes use of these concepts. Broy et al. have explored solutions for using S3D for in-car applications [4].

When evaluating the manufacturer's UI in an off-the-shelf S3D mobile phone, Sunnari et al. [43] found that the S3D design was seen as visually pleasant and entertaining, but lacking in usability, and the participants had difficulties seeing any practical benefit of using stereoscopy. Other than for just hedonistic value, stereoscopy should be incorporated to the mobile UI in a way that it improves not only the visual design, but also the usability [15]. Considering this, more research on S3D for mobile devices equipped with an autostereoscopic display for both user experience and depth perception is still needed.

2.2   Touch Screen Interaction and S3D

In the monoscopic case, the mapping between an on-surface touch point and the intended object point in the virtual scene is straightforward, but with stereoscopic projection this mapping introduces problems [44]. To enable direct 3D "touch" selection of stereoscopically displayed objects in space, 3D tracking technologies can capture a user's hand or finger motions in front of the display surface. Hilliges et al. [21] investigated an extension of the interaction space beyond the touch surface. They tested two depth-sensing approaches to enrich multi-touch interaction on a tabletop setup. Although 3D mid-air touch provides an intuitive interaction technique, touching an intangible object, i.e. touching the void [8], leads to potential confusion and a significant number of overshoot errors. This is due to a combination of three factors: depth perception being less accurate in virtual scenes than in the real world, see e.g. [41], the introduced double vision, and also vergence-accommodation conflicts. Since there are different projections for each eye, the question arises: where do users touch the surface when they try to "touch" a stereoscopic object?

As described by Valkov et al. [47], for objects with negative parallax the user is limited to touch interaction on the area behind the object, since without additional instrumentation touch feedback is only provided at the surface. Therefore the user has to reach through the visual object to reach the touch surface with her finger. If the user reaches into an object while focusing on her finger, the stereoscopic effect for the object will be disturbed, since the user's eyes are not accommodated and converged on the projection screen's surface. Thus the left and right stereoscopic images of the object's projection would appear blurred and could not be merged anymore. However, focusing on the virtual object leads to a disturbance of the stereoscopic perception of the user's finger, since her eyes are converged to the object's 3D position. In both cases touching an object may become ambiguous [47]. To reduce the perception problems associated with reaching through an object to the touch screen surface, Valkov et al. [46] created a prototype where the selected object moved with the user's finger to the screen surface.

In principle, the user may touch anywhere on the surface to select a stereoscopically displayed object. However, in perceptual experiments Valkov et al. [47] found that users actually touch an intermediate point that is located between both projections with a significant offset towards the user's dominant eye. Bruder et al. [7] compared 2D touch and 3D mid-air selection in a Fitts' Law experiment for objects that are projected with varying disparity. Their results show that the 2D touch performs better close to the screen, while 3D selection outperforms 2D touch for targets further away from the screen.

Multi-touch technology provides a rich set of interactions without any instrumentation of the user, but the interaction is often limited to almost zero disparity [40]. Recently, multi-touch devices have been used for controlling the 3D position of a cursor through multiple touch points [1, 42]. These can specify 3D axes or points for indirect object manipulation. Interaction with objects with negative parallax on a multi-touch tabletop setup was addressed by Benko et al.'s balloon selection [1], as well as Strothoff et al.'s triangle cursor [42], which use 2D touch gestures to specify height above the surface. Grossman & Wigdor [14] provided an extensive review of the existing work on interactive surfaces and developed a taxonomy for classification of this research.

Considering the mobile device domain, a comprehensive body of work exists examining the input accuracy of 2D touch screens, e.g. Holtz and Baudisch [22] and Parhi et al. [36]. Holtz summarized that inaccuracy is largely due to a "parallax" artifact between user control based on the top of the finger and sensing based on the bottom side of the finger is particularly relevant, as in the S3D case another on-screen parallax effect is introduced. On small mobile screens touch accuracy is even more critical than on large touch devices. The effect of walking on touch accuracy for 2D touch screen UIs has also been previously researched [3, 25, 37]. For example, Kane [25] found variation in the optimal button size for different individuals, whilst Perry and Hourcade [37] concluded that walking slows down users' interaction but has little direct effect on their touch accuracy.

2.3   Positioning Against Related Work

Taking the strong emphasis on visual design in S3D products and the output orientated prior art, there is a clear need to further investigate the interaction design aspects of S3D UIs. In this chapter, we focus on assessing touch screen interaction with S3D UIs in both large-screen tabletop and small screen mobile formats, in particular considering the target selection accuracy of objects at different depths and positions. Our aim is to provide practical information to assist the designers of S3D UIs.

Additionally, an understanding of the effect of the mobile domain on users' input accuracy when selecting targets is critical for the mobile UI designer. This extends the current body of research which has focused either to evaluate S3D in static conditions, or interaction with a 2D UI whilst on the move; the combination of S3D and user motion is not well researched in the literature.

3   Study I – Tabletop S3D Interaction

In this section, we describe experiments in which we analyzed the touch behavior as well as the precision of 2D touching of 3D objects displayed stereoscopically on a tabletop surface. We used a standard ISO 9241-9 selection task setup on a tabletop surface with 3D targets displayed at different heights above the surface, i.e. with different negative parallaxes. Further details about this study and a comparison to 3D selection can be found in [7]. In this section, we focus on the S3D touch technique, in which subjects have to push their finger through the 3D stereoscopically displayed object (i.e. with negative parallax) until it reaches the 2D touch surface.

3.1   Participants

Ten male and five female subjects (ages 20–35, M = 27.1, heights 158–193 cm, M = 178.3 cm) participated in the experiment. Subjects were students or members of the departments of computer science, media communication or human computer-interaction. Three subjects received class credit for participating in the experiment. All subjects were right-handed. We used the Porta and Dolman tests (see [27]), to determine the sighting dominant eye of subjects. This revealed eight right-eye dominant subjects (7 males, 1 female) and five left-eye dominant subjects (2 males, 3 females). The tests were inconclusive for two subjects (1 male, 1 female), for which the 2 tests indicated conflicting eye dominance. All subjects had normal or corrected to normal vision. One subject wore glasses and four subjects wore contact lenses during the experiment. None of the subjects reported known eye disorders, such as color weaknesses, amblyopia or known stereopsis disruptions. We measured the interpupillary distance (IPD) of each subject before the experiment, which revealed IPDs between 5.8 cm and 7.0 cm (M = 6.4 cm). We used each individual's IPD for stereoscopic display in the experiment. Altogether 14 subjects reported experience with stereoscopic 3D cinema, 14 reported experience with touch screens, and 8 had previously participated in a study involving touch surfaces. Subjects were naive to the experimental conditions. Subjects were allowed to take a break at any time between experiment trials in order to minimize effects of exhaustion or lack of concentration. The total time per subject including pre-questionnaires, instructions, training, experiment, breaks, post-questionnaires, and debriefing was about 1 h.

3.2 Study Design

Materials. For the experiment we used a 62 × 112 cm multi-touch enabled active stereoscopic tabletop setup as described in [7]. The system uses rear diffuse illumination for multi-touch. For this, six high-power infrared (IR) LEDs illuminate the screen from behind. When an object, such as a finger or palm, comes in contact with the diffuse surface it reflects the IR light, which is then sensed by a camera. We used a 1024 × 768 PointGrey Dragonfly2 with a wide-angle lens and a matching IR band-pass filter at 30 frames per second.

We used a modified version of the NUI Group's CCV software to detect touch input on a Mac Mini server. Our setup used a matte diffusing screen with a gain of 1.6 for the stereoscopic back projection. We used a 1280 × 800 Optoma GT720 projector with a wide-angle lens and an active DLP-based shutter at 60 Hz per eye. Subjects indicated target selection using a Razer Nostromo keypad with their non-dominant hand. To enable view-dependant rendering, an optical WorldViz PPT X4 system with sub-millimeter precision and sub-centimeter accuracy was used to track the subject's head in 3D, based on wireless markers attached to the shutter glasses. Additionally, although not reported on in the scope of this chapter, a diffused IR LED on the tip of the index finger of the subject's dominant hand enabled tracking of the finger position in 3D (See [7]).

The visual stimulus consisted of a 30 cm deep box that matches the horizontal dimensions of the tabletop setup (see Fig. 2). The targets in the experiment were represented by spheres, which were arranged in a circle, as illustrated in Fig. 2. A circle consisted of 11 spheres rendered in white, with the active target sphere highlighted in blue. The targets were highlighted in the order specified by ISO 9241-9. The center of each target sphere indicated the exact position where subjects were instructed to touch with their dominant hand in order to select a sphere. The size, distance, and height of target spheres were constant within circles, but varied between circles. Target height was measured as positive height from the screen surface. The virtual scene was rendered on an Intel Core i7 3.40 GHz computer with 8 GB of main memory, and an Nvidia Quadro 4000 graphics card.

Test Procedure. For our experimental analyses and description we used a 5 × 2 × 2 within-subjects design with the method of constant stimuli, in which the target positions and sizes are not related from one circle to the next, but presented randomly and uniformly distributed. The independent variables were target height (between 0 cm and 20 cm, in steps of 5 cm), as well as target distance (16 cm and 25 cm) and target size (2 cm and 3 cm). Each circle represented a different index of difficulty (ID), with combinations of 2 distances and 2 sizes. The ID indicates overall task difficulty [13]. It implies that the smaller and farther a target, the more difficult it is to select quickly and accurately. Our design thus uses four uniformly distributed IDs ranging from

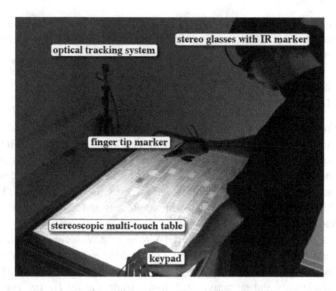

Fig. 2. Experiment setup: photo of a subject during the experiment (with illustrations). As illustrated on the screen, the target objects are arranged in a circle.

approximately 2.85 bps to 3.75 bps, representing an ecologically valuable range of difficulties for such a touch-enabled stereoscopic tabletop setup. As dependent variables we measured the on-display touch areas for 3D target objects.

At the start of the test, subjects were positioned standing in an upright posture in front of the tabletop surface as illustrated in Fig. 2. To improve comparability, we compensated for the different heights of the subjects by adjusting a floor mat below the subject's feet, resulting in an (approximately) uniform eye height of 1.85 cm for each subject during the experiment. The experiment started with task descriptions, which were presented via slides on the tabletop surface to reduce potential experimenter bias. Subjects completed 5 to 15 training trials to ensure that they correctly understood the task and to minimize training effects. Training trials were excluded from the analysis.

In the experiment, subjects were instructed to touch the center of the target spheres as accurately as possible, for which they had as much time as needed. For this, subjects had to push their finger through the 3D sphere until it reached the 2D touch surface. Subjects did not receive feedback whether they "hit" their target, i.e., subjects were free to place their index finger in the real world where they perceived the virtual target to be. We did this to evaluate the often-reported systematical over- or under-estimation of distances in virtual scenes, which can be observed even for short grasping-range distances, as also tested in this experiment. Moreover, we wanted to evaluate the impact of such misperceptions on touch behavior in stereoscopic tabletop setups. We tracked the tip of the index finger. When subjects wanted to register the selection, they had to press a button with their non-dominant hand on the keypad. We recorded a distinct 2D touch position for each target location for each configuration of independent variables.

3.3   Results

In this section we summarize the results from the tabletop S3D touch experiment. We had to exclude two subjects from the analysis who obviously misunderstood the task. We analyzed these results with a repeated measure ANOVA and Tukey multiple comparisons at the 5 % significance level (with Bonferonni correction).

We evaluated the judged 2D touch points on the surface relative to the potential projected target points, i.e., the midpoint (M) between the projections for both eyes, as well as the projection for the dominant (D), and the non-dominant (N) eye. Figure 3 shows scatter plots of the distribution of the touch points from all trials in relation to the projected target centers for the dominant and non-dominant eye for the different heights of 0 cm, 5 cm, 10 cm, 15 cm and 20 cm (bottom to top). We normalized the touch points in such a way that the dominant eye projection D is always shown on the left, and the non-dominant eye projection N is always shown on the right side of the plot. The touch points are displayed relative to the distance between both projections.

As illustrated in Fig. 3, we observed three different behaviors. In particular, eight subjects touched towards the midpoint, i.e., the center between the dominant and non-dominant eye projections. These include the two subjects for whom eye dominance estimates were inconclusive. We arranged these subjects into the group G_M. Furthermore, three subjects touched towards the dominant eye projection D, which we refer to as group G_D, and three subjects touched towards the non-dominant eye projection N, which we refer to as group G_N. This points towards an approximately 50/50 % split in terms of behaviors in the population, i.e. between group G_M and the composite of groups G_D and G_N.

We found a significant main effect of the three groups ($F(2,11) = 71.267$, $p < .001$, partial-eta^2 = .928) on the on-surface touch areas. Furthermore, we found a significant two-way interaction effect of the three groups and target heights ($F(8,44) = 45.251$, $p < .001$, partial-eta^2 = .892) on the on-surface touch areas. The post hoc test revealed that the on-surface target areas significantly vary ($p < .001$) for objects that are displayed at heights of 15 cm or higher. For objects displayed at 10 cm height group G_D and G_N vary significantly ($p < .02$). No significant difference was found for objects displayed below 10 cm. As illustrated in Fig. 3, for these heights the projections for the dominant and non-dominant eye are proximal, and subjects touched almost the same on-screen target areas.

Considering the on-surface touch areas, we found that on average the relative touch point for group GD was 0.97D + 0.03 N for projection points D∈ℝ2 and N∈ℝ2, meaning the subjects in this group touched towards the projection for the dominant eye, but slightly inwards to the center. The relative touch point for group GN was 0.11D + 0.89 N, meaning the subjects in this group touched towards the projection for the non-dominant eye, again with a slight offset towards the center. Finally, for group GM we found that on average the relative touch point for this group was 0.504D + 0.596 N. We could not find any significant difference for the different heights, i.e., the touch behaviors were consistent throughout the tested heights.

However, we observed a trend of target height on the standard deviations of the horizontal distributions (x-axis) of touch points for all groups as shown in Fig. 3. For 0 cm target height we found a mean standard deviation (SD) of 0.29 cm, for 5 cm

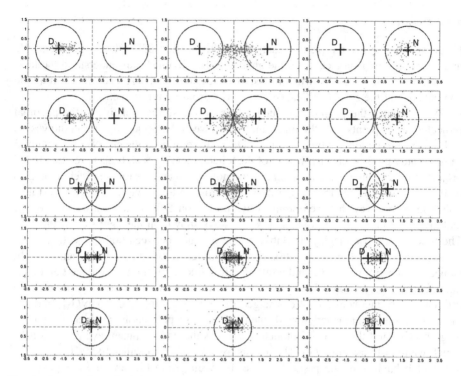

Fig. 3. Scatter plots of relative touch points between the dominant (D) and non-dominant (N) eye projections of the projected target centers on the surface for the 2D touch technique. Black crosses indicate the two projection centers. Black circles indicate the approximate projected target areas for the dominant and non-dominant eye. Top to bottom rows show results for 20 cm, 15 cm, 10 cm, 5 cm, and 0 cm target heights. The left column shows subject behavior for dominant-eye touches (3 subjects), the middle for center-eye touches (8 subjects), and the right for non-dominant-eye touches (3 subjects). Note that the distance between the projection centers depends on the target height.

SD 0.32 cm, for 10 cm SD 0.42 cm, for 15 cm SD 0.52 cm, and for 20 cm SD 0.61 cm. For the vertical distribution (y-axis) of touch points and at 0 cm target height we found a mean SD of 0.20 cm, for 5 cm SD 0.20 cm, for 10 cm SD 0.25 cm, for 15 cm SD 0.29 cm, and for 20 cm SD 0.30 cm.

Minimum Touch Target Sizes. For practical considerations and to evaluate the ecological validity of using the 2D touch technique for selections of targets at a height between 0 cm and 10 cm, we computed the minimal on-surface touch area that supports 95 % correct detection of all 2D touch points in our experiment. Due to the similar distributions of touch points between the three behavior groups for these heights shown in Fig. 3, we determined the average minimal 95 % on-surface region over all participants. Our results show that an elliptical area with horizontal and vertical diameter of 1.64 cm and 1.07 cm with a center in the middle between the two projections is sufficient for 95 % correct detection. This rule-of-thumb heuristic for on-surface target areas is easy to implement and ecologically valuable considering the 'fat finger problem' [20, 30]. Due to this problem, objects require a relatively large size of between 1.05 cm to 2.6 cm for reliable acquisition, even in monoscopic touch-enabled tabletop environments.

4 Study II – Mobile S3D Interaction

To systematically evaluate interaction with S3D in the mobile context, we designed a study to investigate users' performance when interacting with a mobile S3D device and compared this to the 2D case. Here, we examine interaction with both negative and positive parallax (compared to our tabletop study in which only on-screen and negative parallax were considered). In contrast to the tabletop study, the mobile study utilized an autostereoscopic display and hence did not require shutter glasses to be worn to see the stereoscopic effect. Additionally, in this study no viewer dependent adjustment of the rendered view was made. As mobility is by definition a core of this usage context, we also examined differences between interaction when static and walking.

4.1 Participants

The average age of the users (n = 27) was 30.4 years (varying from 10 to 52), and 19 of the users were male and 8 were female. Tests to identify the user's dominant eye (Portas method, see [27]) and stereovision were conducted. Of the 27 users 16 were found to have a dominant right eye whilst 9 had a dominant left eye. Two users were unable to determine which of their eyes was dominant. 24 of the users were right-handed and the remaining 3 left-handed. Most had watched a 3D movie, and approximately half had previously been exposed to a 3D TV or a 3D camera.

4.2 Study Design

As we wanted to focus only on the depth effect due to the stereoscopy, no additional visual depth clues such as shadows, object size or color were used in our test application. In all the tests a background wallpaper image was used, as in pilot tests the use of some type of textured background at a positive disparity/parallax (Fig. 1) was found to improve 3D perception. The background image chosen was a fine mesh pattern at 45°. This was chosen as it gave good 3D perception, but was such that it would not adversely influence the position of the users' presses on targets. In the 2D tests the background was positioned at z-depth = zero and in the 3D tests at z-depth = 10. Table 1 describes the z-depth convention and values used in the study. All the tests were conducted on an autostereoscopic touchscreen mobile device, the LG P920 Optimus 3D mobile phone. This has a 4.3" display with a resolution of 480 × 800 pixels (with 217 ppi) and running Android 2.3.5.

Our target was to investigate the fundamental accuracy of users, hence our method avoided the use of any UI elements such as buttons, whose visuals could influence the position in which users tapped the screen. Thus, we followed a method similar to the crosshairs approach employed by Holtz and Baudisch [22]. Our accuracy test method presented small circular targets, one at a time. The user was instructed to tap on the center of each target with their index finger. For each tap the coordinates of both the press and release touch events were logged, as well as the time between the target being displayed and the user pressing the target.

The accuracy test was conducted in the 2D and S3D condition. In the 2D condition a random sequence of 15 targets was presented. Each target was presented once. The targets were positioned in a grid pattern of 5 horizontal by 3 vertical. This pattern was chosen to give more data points and resolution in the horizontal axis, which is the most interesting for S3D. The targets and background image were displayed as normal 2D objects.

Table 1. Reference stereoscopic depths and calculated apparent distance behind the display (assuming viewing distance of 330 mm and inter-pupil distance of 63 mm)

Accuracy test z-depths	Apparent distance behind screen (mm)
10	12.0
3	3.6
2	2.5
0	0 (on screen)
-1	-1.2 (in front of screen)
-2	-2.5 (in front of screen)

In the S3D version of the accuracy test the same grid of 15 targets was used, but positioned at 5 different z-depths (see Table 1), thus making a total of 75 targets. The 75 targets were presented in a random order of x, y and z position, such that each target was presented once. In the S3D test, the background wallpaper image was positioned at depth $z = 10$, behind the screen. It should be noted that due to the difference in background depth, the S3D test with targets at $z = 0$ was not exactly equivalent to the 2D test.

Based on an eyes-to-screen distance of 330 mm (the mean of 3 pilot users) and an inter-pupil distance of 63 mm (see [12]) the object distances in front of and behind the screen are given in Table 1. These values give an approximation of the apparent depth of objects presented at the different reference depths (the distance of the screen from the eyes was not fixed during the tests).

4.3   Test Procedure

An acclimatization task of looking through an S3D image gallery, provided as default by the device manufacturer, enabled users to get accustomed to the S3D effect. Users were instructed to experiment to find the best position, i.e. distance from device to eyes, viewing angle, etc., where they could best see the 3D effect of the gallery images.

Each of the interaction tests: 2D accuracy and S3D accuracy were conducted in two different conditions: while seated (static) and whilst walking (on the move). For the walking tests, a route consisting of two markers 6 m apart was marked on the laboratory floor. Users were instructed to walk at normal speed around the marked route whilst completing the tests. The order of presentation of the conditions was counterbalanced between participants, to reduce any effect of user learning or other side effects. Users also provided subjective feedback on the comfort of using the interface: this is reported in [9].

4.4 Results

Analysis of the accuracy tests is based on the recorded co-ordinates of the touch press event (i.e. "finger down"), as this defines the users' fundamental accuracy. Presses that were more than 12 mm from the center of the visual target were considered as accidental and excluded from further analysis [31] (Removed data: static 2D = 2, 3D = 4, walking 2D = 3, 3D = 10).

Heatmaps of the press points for each test are shown in Fig. 4, which combines the presses for all of the depth layers. These charts plot the offset of each press point from the center of the presented circular target, i.e. the center of the visual target is at the origin of the chart.

The absolute distance of each press point from the center of the visual target (i.e. ignoring the direction of the error) is presented in Fig. 5. For the S3D layers (i.e. within the groups of five bars in Fig. 5), there was no significant difference between the mean error distance for each of the 3D depth layers (based on ANOVA analysis). Thus, targets at screen level ($z = 0$) were pressed no more accurately than targets at other depths. For all the S3D depth layers combined together, a 2-way ANOVA test revealed

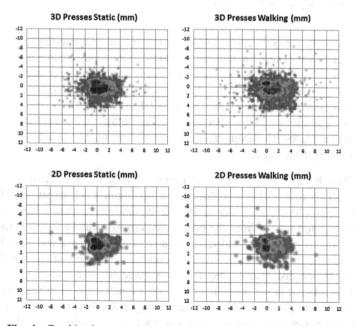

Fig. 4. Combined press points relative to target center for S3D and 2D.

both walking and S3D caused a significant increase in the mean error distance, $(F(2,4841) = 28.8, p < .001)$ and $(F(2,4841) = 54.9, p < .001)$, respectively. There was no significant interaction between the fixed factors.

Comparing the S3D $z = 0$ results, where the target is placed at screen depth, with the corresponding 2D cases, where the target is also at screen depth (circled on Fig. 5), showed significant degradation in accuracy in the S3D cases as compared with 2D cases for $z = 0$ $(T(403) = 3.52, p < .001$ and $T(402) = 4.35, p < .001$ for static and walking cases respectively). This difference is due to the S3D cases having a background image at positive disparity, placing the targets within a stereoscopic scene.

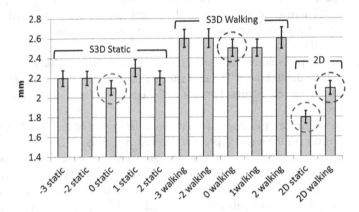

Fig. 5. Mean distance from target center for 2D & individual 3D layers whilst static and walking. Error bars show standard error of mean.

Additionally, in the S3D cases, the $z = 0$ targets are presented in a sequence of targets at other depths. These findings are important as they indicate that from the touch accuracy point of view, there is no benefit in placing targets at screen depth compared to any other depth level. Rather, to maximize touch accuracy, it is recommended that if 3D is not beneficial for other purposes within a UI, it is turned off.

Following a similar method to that employed in Study I, we separated users based on their dominant eyes. However in this case we were unable to ascertain any significance between the groups.

Minimum Touch Target Sizes. One approach to evaluate the degradation in accuracy caused by walking and S3D is to calculate the minimum touch target sizes for each case. The minimum touch target size is the size required for users to reliably hit touch targets in each tested mode. From the press point error distribution we calculated the 98 percentile points in positive and negative directions for both x and y. Taking the approach described by [36], this defines the bounding rectangle that captures 95 % of user presses. As the distributions are slightly skewed to either positive or negative sides, and in practice it is not possible to position a touch target at exactly the correct offset from the center, thus we take double the larger absolute value, which may be considered as the minimum size for targets in the UI. It should be noted that these

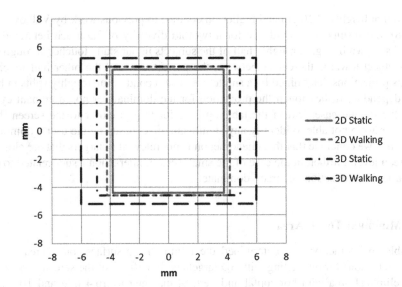

Fig. 6. Minimum target sizes for finger usage (95 % press success).

Table 2. Minimum target sizes for finger usage.

Conditions	Negative 98 percentile (mm)	Positive 98 percentile (mm)	Minimum target size (95 % presses on target) width x height (mm)
2D Static	x:3.5 y:3.9	x:3.8 y:4.4	7.6 × 8.8
2D Walking	x:3.6 y:3.7	x:4.2 y:4.6	8.4 × 9.2
3D Static	x:4.9 y:3.8	x:4.5 y:4.6	9.8 × 9.2
3D Walking	x:6.0 y:4.1	x:5.0 y:5.2	12.0 × 10.4

values are based on index finger usage, as in the test protocol. For the thumb, usage sizes will be somewhat larger (see [36]). The minimum target sizes are shown in Table 2 and diagrammatically in Fig. 6.

5 Discussion

5.1 Touch Performance

Examining our results for the tabletop case, the S3D touch technique appears to have significantly influenced different user groups on the on-surface touch area over the range of tested heights. These on-surface touch areas vary significantly for objects displayed at heights of 10 cm and higher. In contrast to previous work by Valkov et al. [47, 48], our results show evidence for a twofold diversity of 2D touch behaviors of users. As shown in Fig. 3, roughly half of the subjects in our study touched through the virtual object towards the center between the projections, and the other half touched towards projections determined by a single eye. The second group roughly splits in half again depending if they touch the projection for the dominant or non-dominant eye.

In the mobile case, and in the tabletop case for targets closer to the screen than 10 cm, we were not able to identify any significant effect based on the user's dominant eye. Hence we conclude that this phenomenon is not relevant to targets that are close to the screen level. Further, in the case of the small screen, interaction issues related to the general touch interaction accuracy dominate.

5.2   Minimum Touch Area

For tabletop interaction, we determined the minimum on-surface touch area (95 % correct detection) for interacting with 3D targets within 10 cm of the screen surface, to be an elliptical area with horizontal and vertical diameter of 16.4 mm and 10.7 mm with a center in the middle between the two projections. For interaction with the mobile device the corresponding area was a slightly smaller rectangle with dimensions of 9.8 mm × 9.2 mm when static, increasing to 12.0 mm × 10.4 mm when walking. However, further analysis of the differences is not informative, due to the large difference in the depth display capabilities between the mobile device and tabletop (1.2 mm vs. 10 cm and above).

Considering also the 2D performance in the mobile case, and comparing with previous studies, we find close agreement. For example, using their contact area model Holtz and Baudisch [22] report a 7.9 mm target size for 95 % accuracy, compared to our 7.6 mm × 8.8 mm (height × width) for the same accuracy. The aspect ratio information in our results provides interesting additional detail.

5.3   Mobile Context

Direct comparison between our results related to the differences in touch target sizes when walking with prior work on this topic is not straightforward, due to differences in study approaches. The majority of previous work has been based on a UI with varied size button targets, and measuring task completion time and error rates (for example [25]). In contrast, our button free approach aimed to minimize the influence of the visual UI design, and focused on the accuracy of separate tap events. Thus, in our test, accuracy was the only dependent variable.

Our approach enabled us to investigate interaction over the full screen area, and we were able to gain insight into the optimal aspect ratio of touch targets. However, our method is more abstract than actual UI based approaches, and requires some interpretation to transfer it to use in actual designs. It does not take account of other interaction issues e.g. related to Fitts' law [13]. Hence, it is perhaps the case that our method serves best as an initial one, the results of which could be validated by a button and task based method.

Based on our results, for a certain screen size, a S3D UI for use whilst walking should have only 60 % of the touch targets of a correspondingly performing 2D UI (based on area calculations from Table 2). This significantly impacts the design of S3D mobile UIs. The degradation in accuracy from the combined effects of S3D and walking appears to be larger than the sum of each individually (Fig. 6). As may be expected, the main part of this degradation is in the horizontal dimension, and hence related to stereoscopic effects. We speculate that, at least in part, this effect is due to users losing the stereoscopic effect, for example when glancing from the display when walking, which is an expected condition for real world interaction in mobile context (see [35]).

5.4   Cognitive Issues of S3D Interaction

Both tabletop and mobile studies highlighted a large variation in the performance of individuals, which, in the mobile case, was further exaggerated when the users were walking. This large variation makes it difficult to identify statistically significant generic differences between the cases examined without a very large test sample. When designing any user interface it is not wise to design for the average user, without considering the variation of user performance. In the case of S3D UIs this suggests that the use of stereoscopic depth as a standalone informative channel in the UI should be approached cautiously. Clearly, when accessibility for users with restricted capabilities is a consideration, the use of S3D, at least as an informative channel should be avoided.

5.5   Future Work

Learning effects related to long term usage were outside the scope of this study, however this would be an interesting topic for future research. Interestingly, visual examination of the distribution of presses in the mobile case (see Fig. 4) suggests a slight shift to the right whilst walking. This appears similar for both 2D and S3D cases. The reason for this is unknown and would require further study. Possible causes could be differences in the distance between the users' eyes and the screen, or differences in the viewing angle. Such small differences may prove relevant for tasks requiring the pressing of very small targets, such as on screen QWERTY keyboards used for text input, while walking.

6   Conclusions

In this chapter we reported on the evaluation of 2D touch interaction with 3D for scenes on touch-sensitive tabletops and touchscreen mobile devices with stereoscopic displays. We analyzed a technique based on reducing the 3D touch problem to two dimensions by having users "touch through" the stereoscopic impression of 3D objects, resulting in a 2D touch on the display surface.

Tabletop. In the case of tabletops, where the scale is relatively large, we identified two separate classes of user behavior, with one group that touches the center between the projections, where the other touches the projection for the dominant or non-dominant eye. The results show a strong interaction effect between input technique and the stereoscopic parallax of virtual objects.

Mobile Device. For the smaller mobile device S3D touch screen, the mean press positions on visual targets at all z-depths were almost identical, i.e. there was no discernible offset caused by placing targets at different z-depths. However, the variance in press accuracy in the S3D case was much larger than for 2D, and this difference was even more pronounced when walking. When in static usage, to achieve the same performance as 2D, S3D touch targets need to be horizontally wider (7.6 mm for 2D vs. 9.8 mm, for 3D for 95 % presses on target).

To support usage while walking, the size of the touch targets on a touch screen S3D user interface needs to be significantly larger than the corresponding minimum for a 2D interface. In our study the minimum sizes for 95 % presses on target was 8.4 mm × 9.2 mm (width × height) for 2D vs. 12.0 mm × 10.4 mm for 3D. Although these values are based on the test device used, it is expected that they should be generally applicable to other similar S3D devices. For example, the presented minimum touch target sizes in millimeters should serve as initial guidelines for UI designers of other S3D touch screen products intended for mobile use.

The main contributions of this work are:

– We have presented the minimum target size for S3D touch targets for both large screen tabletop and small screen mobile touch screen interaction.
– We identified two separate classes of user behavior when "touching through" stereoscopically displayed objects, on larger scale S3D displays (e.g. tabletops).
– We validated that the 2D touch technique performs well for selection of objects up to about 10 cm height from the display surface.
– For the mobile device case, both walking and S3D caused a significant increase in the mean error distance.

References

1. Benko, H., Feiner, S.: Balloon selection: a multi-finger technique for accurate low-fatigue 3D selection. In: Proceeding IEEE 3DUI, pp. 79–86 (2007)
2. Benzeroual, K., Allison, R.S., Wilcox, L.M.: 3D Display size matters: compensating for the perceptual effects of S3D display scaling. In: Proceedings of CVPRW 2012, pp. 45–52. IEEE (2012)
3. Bergstrom-Lehtovirta, J., Oulasvirta, A., Brewster, S.: The effects of walking speed on target acquisition on a touchscreen interface. In: Proceedings of MobileHCI 2011, ACM (2011)
4. Broy, N., André, E., Schmidt, A.: Is stereoscopic 3D a better choice for information representation in the car? In: Proceedings of the 4th International Conference on Automotive User Interfaces and Interactive Vehicular Applications, pp. 93–100. ACM (2012)

5. Broy, N., Schneegass, S., Alt, F., Schmidt, A.: Framebox and mirrorbox: tools and guidelines to support designers in prototyping interfaces for 3D displays. In: Proceedings of CHI 2014, pp. 2037–2046. ACM (2014)
6. Bruder, G., Steinicke, F., Stuerzlinger, W.: Effects of visual conflicts on 3D selection task performance in stereoscopic display environments. In: Proceedings of 3DUI'10, pp. 115–118. IEEE (2013)
7. Bruder, G., Steinicke, F., Sturzlinger, W.: To touch or not to touch?: comparing 2D touch and 3D mid-air interaction on stereoscopic tabletop surfaces. In: Proceedings of SUI 2013. ACM (2013)
8. Chan, L.W., Kao, H.S., Chen, M.Y., Lee, M.S., Hsu, J., Hung, Y.P.: Touching the void: direct-touch interaction for intangible displays. In: Proceedings CHI 2010, pp. 2625–2634. ACM (2010)
9. Colley, A., Häkkilä, J., Schöning, J., Posti, M.: Investigating mobile stereoscopic 3D touchscreen interaction. In: Proceedings of OzCHI 2013. ACM (2013)
10. Daiber, F., Speicher, M., Gehring, S., Löchtefeld, M., Krüger, A.: Interacting with 3D content on stereoscopic displays. In: Proceedings of PerDis 2014. ACM (2014)
11. Daiber, F., Li, L., Krüger, A.: Designing gestures for mobile 3D gaming. In: Proceedings of MUM 2012. ACM (2012)
12. Dodgson, N.A.: Variation and extrema of human interpupillary distance. Stereoscopic Displays and Applications. In: Proceedings of SPIE 2004 5291, pp. 36–46 (2004)
13. Fitts, P.M.: The information capacity of the human motor system in controlling the amplitude of movement. J. Exp. Psychol. 47, 381–391 (1954)
14. Grossman, T., Wigdor, D.: Going deeper: a taxonomy of 3D on the tabletop. In: Second Annual IEEE International Workshop on Horizontal Interactive Human-Computer Systems, 2007, TABLETOP 2007, pp. 137–144. IEEE (2007)
15. Häkkilä, J., Kytökorpi, K., Karukka, M.: Mobile stereoscopic 3D user experience - calling for user centric design. In: Proceedings of 3DCHI Workshop at CHI 2012 (2012)
16. Häkkilä, J., Posti, M., Schneegass, S., Alt, F., Gultekin, K., Schmidt, A.: Let me catch this! experiencing interactive 3D cinema through collecting content with a mobile phone. In: Proceedings of CHI 2014. ACM (2014)
17. Häkkilä, J., Posti, M., Ventä-Olkkonen, L., Koskenranta, O., Colley, A.: Mobile photo sharing through collaborative space in stereoscopic 3D. In: Proceedings of MUM 2013. ACM (2013)
18. Häkkilä, J., Posti, M., Koskenranta, O., Ventä-Olkkonen, L.: Design and evaluation of mobile phonebook application with stereoscopic 3D user interface. In: CHI 2013 Extended Abstracts on Human Factors in Computing Systems, pp. 1389–1394. ACM (2013)
19. Häkkinen, J., Kawaid, T., Takatalo, J., Mitsuya, R., Nyman, G.: What do people look at when they watch stereoscopic movies? In: Proceedings of SPIE 2010, pp. 7524 (2010)
20. Hall, A.D., Cunningham, J.B., Roache, R.P., Cox, J.W.: Factors affecting performance using touch entry systems: Tactual recognition fields and system accuracy. J. Appl. Psychol. 4, 711–720 (1988)
21. Hilliges, O., Izadi, S., Wilson, A.D., Hodges, S., Garcia-Mendoza, A., Butz, A.: Interactions in the air: adding further depth to interactive tabletops. In: Proceedings of ACM UIST 2009, pp. 139–148 (2009)
22. Holtz, C., Baudisch, P.: Understanding touch. In: Proceedings of CHI 2011, pp. 2501–2510. ACM (2011)
23. Huhtala, J., Karukka, M., Salmimaa, M., Häkkilä, J.: Evaluating depth illusion as method of adding emphasis in autostereoscopic mobile displays. In: Proceedings of MobileHCI 2011. ACM (2011)

24. Jumisko-Pyykkö, S., Dominik Strohmeier, D., Utriainen, T., Kunze, K.: Descriptive quality of experience for mobile 3D video. In: Proceedings of NordiCHI, 2010, pp. 266–275. ACM (2010)
25. Kane, S., Wobbrock, J., Smith, I.: Getting off the treadmill: evaluating walking user interfaces for mobile devices in public spaces. In: Proceedings of MHCI08, pp. 109–118 (2008)
26. Kerber, F., Lessel, P., Mauderer, M., Daiber, F., Oulasvirta, A., Krüger, A.: Is autostereoscopy useful for handheld AR? In: Proceedings of MUM 2013, ACM (2013)
27. Kommerell, G., Schmitt, C., Kromeier, M., Bach, M.: Ocular prevalence versus ocular dominance. Vision. Res. **2003**(43), 1397–1403 (2003)
28. Kooi, F., Toet, A.: Visual comfort of binocular and 3-D displays. Displays **25**(2004), 99–108 (2004)
29. Lambooij, M.T.M., Wijnand, A., IJsselsteijn, W.A., Heynderickx, I.: Visual discomfort in stereoscopic displays: a review. In: Stereoscopic Displays and Virtual Reality Systems XIV, SPIE-IS&T (2007)
30. Loomis, J.M., Knapp, J.M.: Visual perception of egocentric distance in real and virtual environments. In: Hettinger, L.J., Haas, M.W., (eds.) Virtual and Adaptive Environ-ments, volume Virtual and adaptive environments. Mahwah (2003)
31. Matero, J., Colley, A.: Identifying unintentional touches on handheld touch screen devices. In: Proceedings of DIS 2012, pp. 506–509. ACM (2012)
32. Mauderer, M., Conte, S., Nacenta, M.A., Vishwanath, D.: Depth perception with gaze-contingent depth of field. In: Proceedings of CHI 2014, pp. 217–226. ACM (2014)
33. Mikkola, M., Boev, A., Gotchev, A.: Relative importance of depth cues on portable autostereoscopic display. In: Proceedings of MoViD at ACM Multimedia 2010, pp. 63–68 (2010)
34. Mizobuchi, S., et al.: The effect of stereoscopic viewing in a word search task with a layered background. J. Soc. Inform. Display **16**(11), 1105–1113 (2008)
35. Oulasvirta, A., Tamminen, S., Roto, V., Kuorelahti, J.: Interaction in 4-second bursts: the fragmented nature of attentional resources in mobile HCI. In: Proceedings of CHI 2005. ACM (2005)
36. Parhi, P., Karlson, A.K., Bederson, B.B.: Target size study for one-handed thumb use on small touchscreen devices. In: Proceedings of MobileHCI 2006, ACM (2006)
37. Perry, K., Hourcade, J.P.: Evaluating one handed thumb tapping on mobile touchscreen devices. In: Proceedings of Graphics Interface, pp. 57–64 (2008)
38. Posti, M., Ventä-Olkkonen, L., Colley, A., Koskenranta, O., Häkkilä, J.: Exploring gesture based interaction with a layered stereoscopic 3D interface. In: Proceedings of PerDis 2014. ACM (2014)
39. Pölönen, M., Salmimaa, M., Häkkinen, J.: Effect of ambient illumination level on perceived autostereoscopic display quality and depth perception. Displays 32, pp. 135–141. Elsevier (2011)
40. Schöning, J., Steinicke, F., Krüger, A., Hinrichs, K., Valkov, D.: Bimanual interaction with interscopic multi-touch surfaces. In: Gross, T., Gulliksen, J., Kotzé, P., Oestreicher, L., Palanque, P., Prates, R.O., Winckler, M. (eds.) INTERACT 2009. LNCS, vol. 5727, pp. 40–53. Springer, Heidelberg (2009)
41. Sivak, B., MacKenzie, C.L.: Integration of visual information and motor output in reaching and grasping: the contributions of peripheral and central vision. Neuropsychologia **28**(10), 1095–1116 (1990)
42. Strothoff, S., Valkov, D., Hinrichs, K.H.: Triangle cursor: interactions with objects above the tabletop. In: Proceedings of ACM ITS 2011, pp. 111–119 (2011)

43. Sunnari, M., Arhippainen, L., Pakanen, M., Hickey, S.: Studying user experiences of autostereoscopic 3D menu on touch screen mobile device. In: Proceedings of OZCHI 2012, ACM (2012)
44. Teather, R.J., Stuerzlinger, W.: Pointing at 3D targets in a stereo head-tracked virtual environment. In: Proceedings of 3DUI 2011, pp. 87–94. IEEE Computer Society (2011)
45. Teather, R.J., Stuerzlinger, W., Pavlovych, A.: Fishtank fitts: a desktop VR testbed for evaluating 3D pointing techniques. In: CHI 2014 Extended Abstracts on Human Factors in Computing Systems, pp. 519–522. ACM (2014)
46. Valkov, D., Giesler, A., Hinrichs, K.H.: Imperceptible depth shifts for touch interaction with stereoscopic objects. In: Proceedings of the 32nd Annual ACM Conference on Human Factors in Computing Systems, pp. 227–236. ACM (2014)
47. Valkov, D., Steinicke, F., Bruder, G., Hinrichs, K.: 2D Touching of 3D stereoscopic objects. In: Proceedings of of CHI 2011, pp. 1353–1362. ACM (2011)
48. Valkov, D., Steinicke, F., Bruder, G., Hinrichs, K., Schöning, J., Daiber, F., Krüger, A.: Touching floating objects in projection-based virtual reality environments. Proc. EGVE/EuroVR/VEC 2010, 17–24 (2010)
49. Ventä-Olkkonen, L., Posti, M., Häkkilä, J.: How to use 3D in stereoscopic mobile user interfaces – study of initial user perceptions. In: Proceedings of Academic MindTrek 2013. ACM (2013)

Designing for Hover- and Force-Enriched Touch Interaction

Seongkook Heo, Jaehyun Han, and Geehyuk Lee[⊠]

Department of Computer Science, KAIST, Daejeon, Republic of Korea
seongkook@kaist.ac.kr
{jay.jaehyun,geehyuk}@gmail.com

Abstract. As touch-based interfaces become more popular, there are attempts to enhance the touch interface by making the interface more sensitive to the finger. This means that touch surfaces not only sense the location of a finger contact, but also other properties such as a finger hover or the applied force. In this chapter, we summarize the properties of hover- and force-enriched touch and what we should consider to design rich-touch interactions based on the findings from previous works. We present design strategies for rich-touch interactions and example applications, which we developed with the novel touchpad prototype that is capable of measuring a finger hover as well as the finger force applied to the screen. We measured the performance of using rich touch and collected users' feedback through the experiments.

Keywords: Hover touch · Force touch · Interaction design

1 Introduction

Today, many devices, from mobile phones to TV remote controllers, use touch interfaces as their main interface. There has been a large advance in touch interface to support a variety of applications. With the multi-finger sensing capabilities, new gestures such as two-finger pinch, pan, or four-finger close have been introduced [45] for supporting features like zooming, scrolling, or switching between applications [2, 29]. Still, the current touch input does not fully utilize the rich expressivity of our fingers; touch input only detects contact locations of fingers. Yet, when we interact with real-world objects, we automatically control finger postures and force to examine and manipulate objects.

The use of these physical features is common on many modern tablet pen systems. For example, Wacom Intuos Pro [44] tracks the pen while in the air, recognizes tilt angle of the pen, and measures pressure levels. These features have been widely investigated and used to enhance menu selection [10, 34, 35] and to enable a 3D brush model for a more realistic painting simulation [31].

Using physical features of a finger is not as common as the pen interface. It may be due to the bare finger's difficulty to transmit its hover location, posture, or the force we are applying to the screen. Some physical features already became available on commercial products. For example, recent smart phones like Samsung Galaxy S4 [40] can detect a finger hover, and Research In Motion showed a mobile phone [38] that can measure normal force applied to the screen. Synaptics Force Pad [41] measures normal forces independently for multiple finger touches.

© Springer International Publishing Switzerland 2015
T. Wyeld et al. (Eds.): OzCHI 2013, LNCS 8433, pp. 68–87, 2015.
DOI: 10.1007/978-3-319-16940-8_4

However, to our knowledge, there are a few studies on touch interfaces that utilize both finger hover and multi-directional force in combination with touch input. No study suggested how to design interaction for these physically-enriched touch interfaces. In this chapter, we analyze previous studies using rich touch interactions, and derive interaction design guidelines from them. We also describe our implementation of the hover- and force-sensing touchpad prototype and show our design of new interactions for the touchpad. Through user studies, we found that rich touch can be useful for both increasing the input vocabulary and enabling more natural interaction.

2   Rich-Touch Interaction

There has been an intensive research on enriching touch interaction using additional modalities such as hover/proximity and force. In this section, we summarized the previous studies by types of modalities used and whether it is for measuring or using the modality.

2.1   Hover-Enriched Touch

Measuring Hover. Many finger-hover sensing techniques were developed for large touch surfaces. Rekimoto [36] implemented a proximity sensing multi-touch table system and demonstrated the possible scenarios. Z-touch [42] installed multi-layered line laser modules and a camera to detect finger postures and the hand location over the tabletop surface. Medusa by Annett et al. [1] uses 138 infrared (IR) proximity sensors on the bezel of a touch table to detect users' information such as presence, proximity, or which arm they are using. Hilliges et al. [23] implemented a camera-based, proximity sensing, multi-touch table and introduced mid-air gestures. Moreover, the Leap Motion Controller [26] measures the 3D location of the finger over the approximate are of 76 × 30 mm.

For smaller screens, Tsukada and Hoshino [40] added an IR layer on the touch panel so that the system can detect a finger at slightly above the touch panel. Cypress Semiconductor developed True-Touch [9], a capacitive touch screen with proximity-sensing capability. Hirsch et al. [25] implemented a depth-image capturing display by attaching an image sensor at back of an LCD splay and capturing images while changing aperture patterns with display pixels. Choi et al. [6] developed a proximity-sensing touchpad using IR LEDs to show the potential of the remote touch concept. Choi et al. [8] developed ThickPad [7], which is a device with the size of a laptop touchpad capable of sensing proximity images. Gu et al. [11] used a similar optical approach to make a long touchpad that is meant to replace the laptop palm rest.

Using Hover. From existing literature, we obtained three main uses of the finger hover: previewing, adjusting interaction parameters, and mimicking physical interaction. By having an additional state of a touch, hover, users can use the GUI similarly to how they use a 3-state input device like a mouse. A finger hover can be regarded as a tracking state of the mouse to provide a preview of the items that can be shown to a user. RemoteTouch [6] visualizes the position of a finger hovering over the surface to let users know their finger location without seeing the touchpad, in order to provide users a similar experience to using a touch screen. On a Samsung Galaxy S4 smartphone [40], hovering a finger over a selectable item, like a picture album, shows thumbnail previews of the pictures in the album. Users may also locate their fingers above the calendar item to preview schedules for the day underneath the finger. Tsukada and Hoshino [43] showed that the finger hover over an item may trigger a balloon help to pop up. Marquardt et al. [30] showed the use of a finger hover to reveal available interactions underneath the finger as an example scenario. Medusa [1] detects whether the user's hand is in the pre-touch stage and shows a gesture guide.

Another use of a finger hover is to change interaction parameters. Han and Park [14] developed a method of using finger hover to change the scale factor. They compared touch-based, hover-based, and a hybrid method using both touch and hover zooming. Result of the comparison test showed that the hover-based zoom technique was both the fastest and the most preferred. Takeoda et al. [42] presented a map application that changes zoom levels by the height of the finger. Chun Yu et al. [48] used the finger height information to change the C-D ratio of a finger to enable multiscale navigation. Marquardt et al. [30] showed that the height of the finger movement may change the precision for cases like scaling.

When a system can detect the hand shape or the hand posture, hover operations can be richer. Choi et al. [8] designed a set of area gestures that utilizes different shapes of the hand over the surface of the ThickPad [7] prototype. LongPad [11] utilizes the proximity image to prevent accidental touches that can be made as the prototype replaces the laptop palm rest.

Hilliges et al. [23] developed a new tabletop computer that uses a holoscreen to switch transparency of the surface to alternatively display user interface and capture the depth image of the hands over the surface. The authors made gestures like grasping with fingers to mimic real-world object manipulation.

2.2   Force-Enriched Touch

Measuring Normal Force. In the real world, we control normal force to move or deform an object. Many studies on using normal force measure the same with force sensors, such as strain gauges and force sensitive resistors (FSR). In 1977, Herot and Weinzapfel [22] presented a force-sensing touch screen with four multi-dimensional strain gauges. Minsky [32] developed a set of gestures that use both touch and force for a force-sensitive touch screen made with four strain gauges at each screen corner. GraspZoom [33] and ForceDrag [19] placed a FSR at the back of the mobile device to provide an additional input dimension to that of the touch input. Some research tried to estimate the normal force without using force sensors. Gu et al. [11] developed an optical touchpad that has a compliant silicon sheet on an array of IR LEDs and photodiodes and that measured the normal force with the optical image made by the deformation of the compliant surface. Benko et al. [5] presented a method that senses force applied with a finger through the change of muscle tensions by utilizing forearm electromyography. Benko et al. [4] introduced the SimPress method, which estimates the surface press of a finger with a change of the finger contact. Recently, research has been done on enabling multi-touch force sensing. Synaptics [41] introduced a pressure-sensitive multi-touch pad that can independently measure the normal force for each finger. Rosenberg and Perlin [39] developed UnMousePad, which measures the multi-touch force with a new principle called interpolating force sensitive resistance. Some research focuses on measuring the tapping force without using force sensors. Hinckley and Song [24] showed that the hard tap can be sensed with inertial sensors. Heo and Lee [17] showed that the built-in accelerometer could be used to discriminate light tap and strong tap by measuring the device movement with a high recognition ratio (>90 %).

Using Normal Force. It is natural to control pressure while writing or drawing with a pen and the pressure has been utilized to enrich GUI operations [34, 35]. However, utilizing normal force of the finger is not as simple as it is with pens, for stronger normal force makes fingers harder to move due to friction. Therefore, many studies using normal force avoid finger movement while applying force. Miyaki and Rekimoto [33] developed GraspZoom, which uses the normal force applied to the mobile device display to change zoom levels and scroll contents. Because the normal force information only has its intensity, the authors used a small sliding gesture to toggle zoom or scrolling directions. Normal force was also used to enable multi-modes touch operations. Heo and Lee [19] used normal force to switch between different touch operations. The authors introduced the force lock concept, which divides the force-based mode selection phase from the touch manipulation phase, resembling the concept of gesture registration and relaxation [47] to ease force interaction. Rendl et al. [37] used Synaptics ForcePad [41] and introduced force-augmented multi-touch gestures. In the paper by Gu et al. [11], the authors showed a piano application that changes the velocity by the strength of the normal force applied to the touch surface.

Measuring and Using Tangential Force. Tangential force, which is force applied perpendicular to the surface, has recently been studied. Heo and Lee [18] developed a force-sensing touch prototype capable of measuring multi-dimensional force. They also designed force gestures that enrich touch gestures with normal and tangential forces and implemented the gestures in applications. Through a user study, the authors found that users prefer the use of tangential force for rate-controlled page flipping, and that having continuous visual feedback on the direction and the amount of tangential force is essential for a better user experience. Harrison and Hudson [16] developed a shear platform that has a touch panel on the display, connected with two self-centering analog joysticks. They also introduced possible scenarios using tangential force. According to Lee et al. [27], tangential force can be used to select one of multiple targets, while the index of difficulty follows Fitts' law, regarding the force as the movement. Recently, Heo and Lee [20] proposed a new method of sensing tangential force for multiple fingers independently by measuring the finger contact displacement made by the finger deformation while applying force.

3   Designing Rich-Touch Interaction

In this section, we summarize the considerations for hover- and force-enriched touch interface, and show possible design strategies of designing rich-touch interaction.

3.1   Considerations for Hover-Enriched Touch Interaction

One interesting factor from the literature review was the effect of the hover interaction on the content. In many cases, hover input was used for functionalities that do not affect the main context. In RemoteTouch [6] and LongPad [11], hover input was only used to indicate the finger location. Other studies utilized the hover input to show additional information about the item under the finger location [1, 30, 40, 43] or to change the zoom scale [14, 30, 42, 48].

When the system is capable of detecting finger postures, detecting a finger hover was sometimes used to affect virtual objects [23, 30, 43, 46]. By detecting the finger shape and posture, they could use finger hover both to show cursor location and to modify the virtual model. The distinct postures were similar to the real-world manipulation: picking objects with pinch gestures and move/manipulate with the hand movement [23, 30], moving objects with swiping gesture towards the other hand [46], or sticking an object to the fingertip [43].

When designing hover-enriched touch interaction, we need to choose whether the hover will affect the content or not. To allow hover to make changes, we need a method to confirm changes, since the hover movement can be made unintentionally before and after touch operations. We may sense posture with hover images or delimit with start/ end touches. Another factor we need to consider is how to map the input. Positional input may fit better for hover input than rate-control input, since the previous study by Zhai [49] showed that the position control performs better than rate control for isotonic input devices.

Other factors we need to take care are the usability factors of performance and fatigue. Previous research [3] showed that the pointing performance in the air is lower than that on a touch screen. Because of that reason, we may use hover movements for vague operations like an area magnification or less explicit operations like previewing. We may increase the Control-Display ratio by using a wider interaction area in the air than the surface for hover operations [30] to overcome a low-precision problem.

3.2   Considerations for Force-Enriched Touch Interaction

Applying force to modify contents is natural. We hold, move, rotate, or deform an object by controlling normal and tangential force. Different from hover input, applying force is an explicit operation; thus it fits more for manipulative operations than previewing. Force input can be used either for discrete input or continuous input for its continu- ousness. Synaptics ForcePad [41], BlackBerry Storm [38], and the prototype used in Force Gestures [18] measure pressure and use it to enable an additional mode: pressed. In ForceDrag [19], the authors used pressure to select the touch operation mode among four modes. Presstures [37] used the pressure to enable pressure-gesture mode.

When force is utilized to control parameters, the force strength is used as a con- tinuous value. In GraspZoom [33], normal force was used to scroll or scale the content. Tangential force was used to turn through pages by varying speed [18] and continuous pan and zoom [16, 20].

Continuous visual feedback is essential for force-enhanced input both for discrete and continuous use. Ramos et al. [34] found that the continuous visual feedback is necessary for pen-based pressure widgets. Heo and Lee [18] described that the users could not use pivot gesture for turning pages, which is a gesture of pushing with tangential force, without continuous visual feedback indicating the direction and strength of the tangential force. The continuous feedback does not only show the current pressure level but also indicates that the system is responsive.

Friction between the finger and the surface increases as more normal force is applied to the surface. Previous work [19, 37] tried to avoid this problem by dividing the gesture registration phase from the performing phase, so that users can perform force-augmented gestures without applying normal force. Since friction can be useful for applying tangential force to the surface, we need to avoid letting users slide their fingers on a surface with pressure.

4   Proximity- and Force-Sensing Touchpad

We used the novel touchpad capable of sensing both finger hover and multi-dimensional force as suggested by Heo et al. [21] to show design examples of a rich-touch. The prototype measures the finger proximity using infrared LEDs and sensors, and measures normal and tangential force with multiple force sensors.

4.1   Hardware Configuration

The hover and force tracking touchpad hardware prototype is shown in Fig. 1. Size of the prototype is 142 × 88 × 21 mm and its interaction space is 113 × 63 × 15 mm, for width, height, and depth, respectively. The prototype consists of three parts including the proximity and touch imaging part, force-sensing part, and sensor data acquisition and transmission part. We also placed a 3 mm-thick acrylic layer on four normal force sensors installed on the circuit board with a 3 mm-thick silicon sheet on the acrylic layer to enhance touch detection. The cover and the housing were 3D printed.

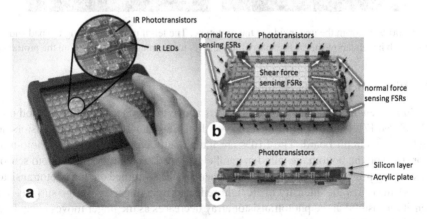

Fig. 1. (a) Prototype device in use with a finger, (b) Prototype device without cover, and (c) Side view of the prototype without cover.

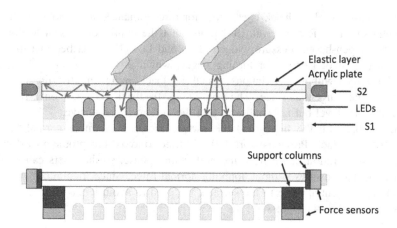

Fig. 2. Optical proximity and touch detection (top) and multi-dimensional force sensing (bottom).

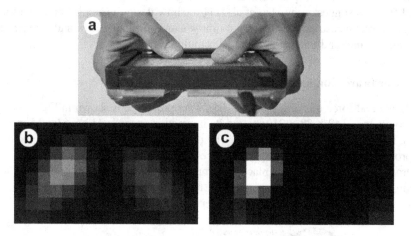

Fig. 3. (a) View from the upper side of the prototype. The left thumb is touching the pad and the right thumb hovers above it. (b) Raw proximity and (c) touch images obtained from the prototype.

4.2 Rich-Touch Sensing Method

Proximity Sensing. The device tracks fingers above the surface using the method used in ThickPad [7]. As shown in Fig. 1, 15 × 8 IR LEDs and 16 × 9 phototransistors are attached to a circuit board. The LEDs are individually controlled and the photo-transistors (S1) on the bottom are wired in parallel and work as a large, single photo sensor. The working principle is shown in Fig. 2. When an LED is turned on, the phototransistor array S1 measures the intensity of IR light reflected by a finger above the surface. The intensity measured at the phototransistor array decreases as the finger moves away from the surface and this relation provides the approximate distance between the finger and the surface. After the measurement process is sequentially repeated for all LEDs, we can get a hover image with 15 × 8 pixels. The obtained hover image is shown in Fig. 3b. The prototype can track fingers at up to 15 mm above the surface.

We apply several image processing steps on raw sensor images to find fingertips (See Fig. 4). The first step is the ambient light compensation. Even if we reduce the effect of the ambient light in the signal processing, the prototype structure itself, such as the acrylic plate and the elastic layer, reflect light. This reflection appears constantly, thus we can compensate for this reflection by simply subtracting the baseline image, which was captured without any fingers, from the raw images. Then, we normalize each pixel value to compensate for the differences between the LEDs and the photo-transistors. We perform a gamma correction and a cubic interpolation to the normalized images to get clearer and smoother images. We then calculate the finger center by calculating the center of mass of the pixels in each finger blob and stabilize the finger position using speed-dependent low-pass filter [28].

Touch Detection. ThickPad [7] used a conductive layer on the touch surface to detect whether a finger is touched or not. However, because the conductive layer is one large sheet made with a transparent conductive film, it can only detect one contact of a finger. To detect multiple finger contacts, we used an optical touch detection method [12], which uses internal scattering of the infrared light. Similar to the proximity sensing method, this method requires an array of LEDs under the acrylic layer. However, this method uses phototransistors located at the side of the acrylic layer. As shown in Fig. 1, 22 phototransistors (S2) were installed at the edges of the acrylic plate for touch detection. These phototransistors (S2) were wired in parallel. When a finger touches the surface, IR light emitted from the LEDs is scattered at the point of contact and reflected inside the elastic acrylic plate to the S2 phototransistors. The phototransistor array S2 measures the intensity of the reflected IR light.

We obtain two sets of sensor values from the phototransistor arrays S1 and S2. Even if we capture two images, each LED is turned on/off once per image frame, because the position of an image pixel is dependent only on the LED states and no interference exists between S1 and S2. To reduce the effect of the ambient light, our prototype reads sensor values twice, when one LED is turned on and when all LEDs are turned off. Then, the prototype subtracts values obtained when LEDs are off from those when LEDs are on; the difference represents the intensity of light reflected from the fingers. For all the signal process steps, the frame rate of the current prototype is at about 50fps for two 15 × 8 pixels images. From these images, we can not only obtain the position of the fingertip, but also the shape and size of the finger contact.

Force Sensing. The touchpad prototype measures force with 2.5 degrees of freedom by measuring normal and tangential force with 12 FSRs. Four FSRs were attached to the circuit board and under the acrylic plate, and two FSRs were attached to each of the four sidewalls along the edges of the acrylic plate (Fig. 1). The FSR used in the prototype is Interlink Electronics FSR^{TM} 400, which can measure force from 0.1 N to 10 N, thus the force sensing range is 0.4 N to 40 N for normal force and 0.2 N to 20 N for tangential force.

Even though there is no finger touching the device, sensor values change with the orientation change of the touchpad, because of the weight of the acrylic plate. In order to reduce this effect of gravity, the system calibrates force sensors by recording the

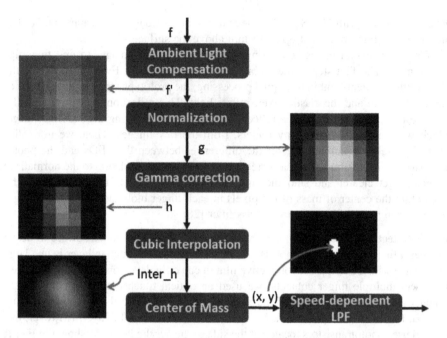

Fig. 4. Fingertip localization process.

sensor values at the moment of initial finger contact and by subtracting those values from sensor values measured while the finger is touching the pad. When calibrated sensor values are collected, the system calculates the force vector. The 2.5D force vector (x, y, z) can be calculated as follows:

$$(F_R - F_L, F_T - F_B, -F_V)$$

Where F_T, F_B, F_L, and F_R are the forces obtained from the top, bottom, left, and right side walls, respectively, and F_V is the sum of the force values under the acrylic plate.

5 Application Scenarios

The proposed touchpad prototype enables hover- and force-enhanced touch interactions. We show possible application scenarios using interaction mappings designed with two different design strategies: (1) multi-level user interaction and (2) mimicry of physical manipulation.

5.1 Design Strategies for Rich-Touch Interaction

We built two design strategies for designing rich-touch interaction with the new touchpad. Those design strategies have different goals: expanding GUI operations and enabling realistic interactions.

Multi-level User Interaction. Using rich information of touch can expand touch user interfaces into multi-level input. We can use the finger hover to facilitate preview and force input that enables rate-controlled directional input with the existing touch input.

Finger hover is temporary and imprecise. We can use the finger hover as a mouse tracking, which is currently missing on touch interfaces. Different from mouse tracking, finger hover has an additional value: height. Users may preview the underlying content while varying the level of detail, like the proximity-based zooming by Harrison and Dey [15]. Touch operations can be used similarly. However, with the support of the hover preview, indirect touch surfaces like a touchpad can be used as an absolute-mapped touch screen such as the RemoteTouch [6].

Users can apply force while using touch operations. Force input is multi-dimensional and isometric. For a clearer GUI operation, we can divide the multi-dimensional force input into five directional force inputs: press and push to four tangential directions. Press can be used as button and pushes can be used to continuously change values of the control at the touch location.

Mimic Physical Interaction. When modeling clay, we utilize our innate ability to control finger postures and force with high precision. We press the clay with fingers to change its shape. We push the clay from the side, or pinch it with the fingers to gather more clay. While we are doing these processes, we naturally move fingers, change the finger angle to control the contact size, or change the force direction and strength. The second interaction design strategy is mimicking the physical manipulation on the computer.

Measured height and location of the finger is used to indicate users where their finger is in the virtual space. The finger location may be shown as a cursor; change of the cursor opacity and shape can be used to indicate the height as used in RemoteTouch [6]. Our touchpad prototype can measure shape and area of the finger contact; those properties also affect the cursor size. In this strategy, touch and force input are not strictly separated and the area of finger contact is also fully utilized. Touch input with different force directions and strengths directly affects the virtual object in the way we interact with real-world objects. The object can be moved, deformed, or rotated with the force, and the contact area determines how large the area affected by the touch operation should be.

5.2 Application 1: Video Browsing

We chose the video browsing application for a multi-level user interaction strategy because the video browsing task needs various operations including skimming a video, scene browsing, and playback speed control. With this application, we aim to provide rich video browsing without adding GUI buttons. The 1-to-1 mapped position control will allow users to quickly scan the video, touch operations will enable quick scene browsing, and rate-controlled force control will enable playback speed control.

The hover operation is the most temporary and unstable positional operation among the three types of operations, because the users can hold their finger in the air without intending any action. We mapped the horizontal axis of the touchpad to the video timeline and showed a preview of the scene corresponding to the touch position

(Fig. 5). Hover preview is used only to skim the video content and to perform the video playback, as it is not a stable operation. The touch operation is more stable than the hover operation. Therefore, we mapped the touch movement to actually change scenes in a playing window. Similar to the hover operation, the horizontal position of a touch is directly mapped to the video timeline.

Rate-control input devices, like a jog shuttle, are widely used for browsing and editing videos. We use tangential force, in place of a rate-control joystick, to support precise video browsing. Users can change the direction and the speed of the video playback by changing the direction and the strength of the tangential force applied to the surface. Because the tangential force input is isometric and continuous, users do not need to move their finger repeatedly to browse through a long stretch of the video. When users are applying tangential force for rate control, an arrow-shaped indicator is displayed on the timeline to show the direction and the speed, as shown in Fig. 5b.

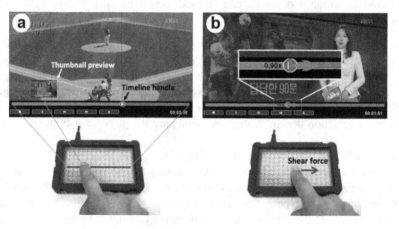

Fig. 5. Interaction mapping for the proximity– and force–sensing touchpad. (a) Horizontal position on the touchpad is directly mapped to the time position on the video timeline, and (b) tangential force on the touchpad changes the playback speed in proportion to the force strength.

5.3   Application 2: 3D Surface Modeling

We developed a 3D surface modeling application with the metaphor of an elastic mesh modeling to mimic physical manipulation. The application fully utilizes the sensing capabilities of the touchpad prototype. Hover and touch sensing capabilities show the finger location and contact size, and change the size of the deformation area. Multidimensional force deforms the surface in both normal and tangential directions. A similar technique has been proposed by Han et al. [13] with an elastic force-sensing structure. Because it could not sense finger hover and tangential force, they used a repoussé and chasing metaphor, which only requires deformation in normal direction.

The application shows the position and contact size of a finger with a circular-shaped cursor by varying its radius. As shown in Fig. 6a, when two thumbs are touching the surface with different postures, two cursors appear with different radiuses.

The different contact size determines how the surface deforms. Applying force with a wide finger pad makes a smooth and large deformation, and pressing with a sharp fingertip makes a sharp deformation at a small area (see Fig. 6b, c). The force not only deforms the surface in the normal direction, it also makes it bend into the tangential direction as shown in the Fig. 6d since the force vector is 2.5 degrees of freedom. The amount of deformation is determined by the magnitude of the force vector.

In our prototype, light touch (<0.4 N in normal, <0.2 N in tangential) does not modify the surface since the FSRs used in the prototype can only measure force stronger than 0.1 N. This prevents an accidental touch to modify the object. Our prototype is limited to 2.5 degrees of freedom; therefore, interactions in the reverse direction cannot be supported. As a solution, we provide a modifier key to invert the direction of the finger movement.

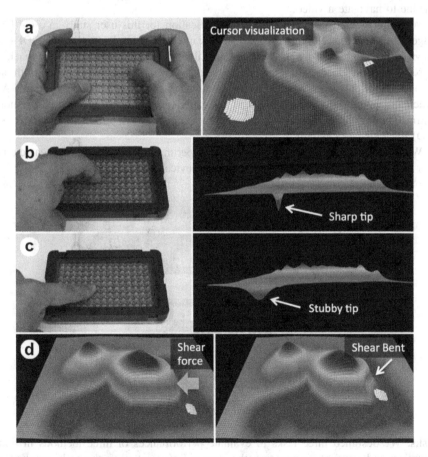

Fig. 6. 3D manipulation mimicking physical interaction. (a) Cursor visualization, (b) precise and sharp manipulation using the fingertip, (c) large and stubby manipulation using the finger pad, and (d) pushing the sidewall using tangential force.

6   Evaluation

6.1   Video Browsing

We conducted a user study to check the usability of the new video interaction made based on our new design strategy. In the user study, we compared three methods: traditional button interface, traditional touchpad method, and proximity- and force-sensing video interaction. For the traditional button interface, we used a Samsung TV remote shown in Fig. 7. The remote control has only two buttons for changing play-back speed (0.25x–8x); we modified the two buttons on the upper side for a 10 s scene jump. For the traditional touchpad method, we used the prototype device developed in the study. Instead of using all sensing capabilities, traditional touchpad methods only use touch detections and presses. Users can move a cursor with the touchpad the way they are using a laptop touchpad and select GUI buttons to jump 10 s backward and forward, locate a cursor on a timeline to see a thumbnail preview, and drag on the timeline to navigate a video.

We implemented the video browsing application for this user study in a C# lan-guage on a Microsoft .Net framework 4.5. User interface is implemented with GDI + as shown in the Fig. 5 and we used VideoLan Library [19] to control video playback.

Nine undergraduate and graduate students (seven male, two female) with an average age of 24 years were recruited through a school online bulletin board. All participants were right-handed and had no experience of using a touch-pad remote control. The study took 30 to 40 min, and participants received 5000 KRW (approx. $5 USD) for their participation.

We used a 46-inch TV as a display for the experiment. Participants were seated on a chair at 2 m away from the TV, and used the device on a table.

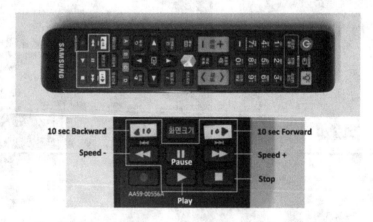

Fig. 7. Samsung TV Remote Controller used in the experiment.

Tasks. We designed three tasks to evaluate performances of three methods in both skimming and precisely navigating the video. Before starting the task, we briefly instructed participants about how to use the controllers. We used six video clips containing sports news briefings, and the videos are 4:01 (min:s), 1:57, 4:02, 1:35,

2:33, and 2:52. Each video clip contains 6 to 14 news briefings of five to 90 s. The order of devices was counterbalanced across participants using the latin-square method.

The first task aimed to measure the performance of skimming the video. For the first task, participants were asked to count news briefings in a video using three methods. They were asked to count as fast as possible but to be aware of not missing any news. Task completion time and the number of missing news briefings were recorded.

The second task measured the video browsing performance. In this task, participants were given papers with a scene of news printed on it. They were asked to find the news briefing with the topic. Because the title of the news is displayed most of the time during the briefings, this task could be performed with a rough navigation.

The third task measured the performance for precise navigation. Participants were asked to find a specific time position in a video. The time position was given in seconds, and participants were asked to match to one second.

Result. Figure 8 shows the average task completion time of the three tasks. The new method was the fastest for all tasks. For the first task, there was a significant difference among three devices ($F(2,34) = 7.033$, $p < 0.01$, repeated-measures ANOVA).

We also conducted paired t-test to figure out the performance difference between two methods and found that the touchpad was significantly slower than the other two methods ($p = 0.049$ for the button remote, $p < 0.001$ for the new touchpad). For the second task and third task, there was no significant difference among the three methods. However, the touchpad and the new touchpad methods showed a near significant difference ($p = 0.051$) on the second task and showed a significant difference ($p = 0.023$) on the third task.

The ratio of the miscounted number over the total number of news briefings for the first task was 1.85 % (three occurrences among 162 briefings) with the touchpad and 0 % with traditional remote control and the new touchpad. We guess the miscount occurred while participants dragged the timeline slider using the touchpad by missing the timeline control cursor.

Table 1 shows the participants' answers on the most and the least preferred method for three tasks. The new touchpad was mostly preferred for the first task with slightly more votes than the others. Participants who preferred the new touchpad answered that the 1-to-1 mapping of the touchpad and the timeline was intuitive and the hover preview was easy to use. For the second task, all but one participant answered that they prefer to use the new touchpad. They answered that the new touchpad enabled instant access to the target news. For the third task, using both the button remote control and the new touchpad received the most votes. Those who preferred the button remote answered that they could comfortably access the target time with the 10-second jump button. One participant answered that he preferred to use the button remote, but it would not be preferred for long video clips.

6.2 3D Surface Modeling

To investigate the user experience of the 3D surface modeling application, we conducted a user study. We implemented the 3D surface modeling application in Processing using OpenGL. As shown in Fig. 6, there is no GUI buttons on a screen and

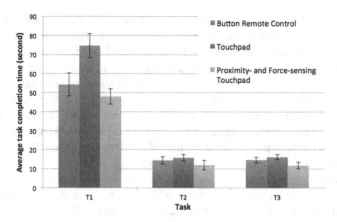

Fig. 8. Average task completion time of three tasks for three methods. Error bars indicate standard errors.

Table 1. Number of participants who prefer /not prefer each method for three tasks.

	Button Remote Control (most / least preferred)	Touchpad	Proximity- and Force-sensing Touchpad
T1	2 / 3*	3 / 3*	4* / 3*
T2	0 / 5*	1 / 4	8* / 0
T3	4* / 2	1 / 6*	4* / 1

*Largest number of each task.

participants can deform the surface by applying multi-dimensional force the touchpad. The normal force direction is inversed when the user presses a modifier key. Participants can rotate the view with arrow keys to better understand the terrain model.

Figure 9 shows the experiment setting. Participants completed the tasks while seated in front of a laptop with 15-inch display. We placed a keypad with a modifier key that changes the deformation direction. Five students (three male, two female) with an average age of 22.4 participated in this study. No participant had experiences in using 3D modeling software, such as Autodesk 3 ds Max or Autodesk Maya. All participants could finish the study in 30 min, and they received 5000 KRW for their participation.

Tasks. Before asking participants to perform tasks, we briefly instructed and demonstrated how to use this touchpad for modeling 3D surfaces and let the user test it. Then we asked participants to make a 3D terrain model following the given pictures using the prototype. The pictures included three different types of terrain: A smooth terrain with several hills, a terrain with many mountains around a large lake, and an island with a cliff. We also asked participants to modify the terrain such as: "please make a river flowing from the lake to the plain." After the experiment, we interviewed the participants about their experience including a 5-point Likert scale questionnaire on 'it was easy to learn', 'it was effective to make 3D terrain models'.

Fig. 9. Experiment setting for 3D surface modeling application

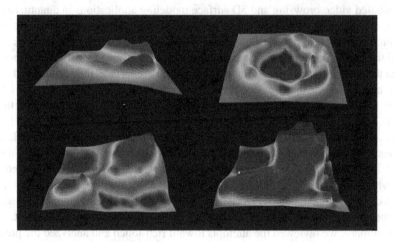

Fig. 10. 3D Terrain models created by the participants.

Result. Figure 10 shows some of the models participants created. All participants could make the terrain models within 10 minutes for each model. Participants usually started by making large structures with their finger pad and then used the fingertip to make sharp details. In the questionnaire, the participants rated that the interface could be easily learned (average answer 4.6 /5) and was effective to make the terrain models (4.2/5).

Overall, the participants were satisfied with the new interface. They answered that the modeling process was easy, intuitive, and fun. They could control the modification range by varying finger posture and modification speed by controlling force strength without reporting any difficulty. Two participants commented that the modeling procedure was like playing with clay or dough and children would have fun with this interface.

Feedback varied on the absolute mapping. Two participants commented that the absolute mapping was confusing since it is different from other touchpads. Two other participants told that it felt strange at first, but they grew to like the mapping because it felt more natural. One participant answered that he had no difficulties. We also asked about the use of the additional direction-changing button and the participants answered that it was okay to use the button. One participant answered that it would be more interesting if the touchpad was double-sided so that one could apply force in any direction without pressing a key.

There have been suggestions on the touchpad size. One participant commented that he wished that the touchpad would be larger so that he could modify the model more precisely. Another participant told that 'if the touchpad is large enough to place the whole hand on it, it would be fun to use the entire hand to make hand prints like we do with clay'.

7   Discussion and Conclusion

We evaluated video browsing and 3D surface modeling applications in quantitative and qualitative ways, respectively, as the goals of the two applications were different. In the video browsing experiment, we showed the potential of using rich touch information to expand the input vocabulary by increasing the input performance. Participants not only showed the shortest task completion time with the new interface, they preferred the rich-touch touchpad for the most for two tasks. Participants especially liked the 1-to-1 hover preview feature, because they could instantly skim the video without changing the use context. The 3D surface modeling application showed that mimicking the physical interaction might be a good way to enable easy and intuitive interaction. Participants could learn how to use the interface within several minutes and they could build models without long tutorial and practice. Both design strategies successfully achieved their goals, which are supporting a layered and more input vocabulary and providing a real-world-like surface manipulation. Designers may choose one of the example strategies or create their own strategy while considering the properties of each type of input.

In summary, we reviewed the attempts toward rich-touch and analyzed the previous works to understand properties of hover-enriched and force-enriched touch interaction. Based on the properties of rich-touch interaction, we pulled out strategies to design interactions for rich-touch interfaces. We implemented a novel touchpad that can measure hover, touch, and multi-dimensional force and two rich-touch enabled applications. With the applications designed with our design strategies, we could find that both the uses of rich touch for multi-level user interaction and mimicking physical interaction were effective.

References

1. Annett, M., Grossman, T., Wigdor, D., Fitzmaurice G.: Medusa: a proximity-aware multi-touch tabletop. In: Proceedings of the UIST 2011, pp. 337–346. ACM (2011)
2. Apple, Magic Trackpad. http://www.apple.com/magictrackpad/
3. Banerjee, A., Burstyn, J., Girouard, A., Vertegaal, R.: Pointable: an in-air pointing technique to manipulate out-of-reach targets on tabletops. In: Proceedings of the ITS 2011, pp. 11–20. ACM Press (2011)
4. Benko, H., Wilson, A.D., Baudisch, P.: Precise selection techniques for multi-touch screens. In: Proceedings of the CHI 2006, pp. 1263–1272. ACM Press (2006)
5. Benko, H., Saponas, T.S., Morris, D., Tan, D.: Enhancing input on and above the interactive surface with muscle sensing. In: Proceedings of the ITS 2009, pp. 93–100. ACM Press. (2009)
6. Choi, S., Han, J., Lee, G., Lee, N., Lee, W.: RemoteTouch: touch-screen-like interaction in the TV viewing environment. In: Proceedings of the CHI 2011, pp. 393–402. ACM Press (2011)
7. Choi, S., Han, J., Kim, S., Heo, S., Lee, G.: ThickPad: a hover-tracking touchpad for a laptop. In: Adjunct. Proceedings of the UIST 2011, pp. 15–16. ACM Press (2011)
8. Choi, S., Gu, J., Han, J., Lee, G.: Area gestures for a laptop computer enabled by a hover-tracking touchpad. In: Proceedings of the APCHI 2012, pp. 119–124. ACM Press (2012)
9. Cypress Semiconductor, TrueTouch. http://www.cypress.com/touch/
10. Grossman, T., Hinckley, K., Baudisch, P., Agrawala, M., Balakrishnan, R.: Hover widgets: using the tracking state to extend the capabilities of pen-operated devices. In: Proceedings of the CHI 2006, pp. 861–870. ACM Press (2006)
11. Gu, J., Heo, S., Han, J., Kim, S., Lee, G.: LongPad: a touchpad using the entire area below the keyboard of a laptop computer. In: Proceedings of the CHI 2013, pp. 1421–1430. ACM Press (2013)
12. Han, J., Choi, S., Heo, S., Lee, G.: Optical touch sensing based on internal scattering in touch surface. Electron. Lett. **48**(22), 1420–1422 (2012)
13. Han, J., Gu, J., Lee, G.: Trampoline: a double-sided elastic touch device for creating reliefs. In: Proceedings of the UIST 2014, pp. 383–388. ACM Press (2014)
14. Han, S., Park, J.: A study on touch & hover based interaction for zooming. In: Extended Abstracts, Proceedings of the CHI 2012, pp. 2183–2188. ACM Press (2012)
15. Harrison, C., Dey, A.K.: Lean and zoom: proximity-aware user interface and content magnification. In: Proceedings of the CHI 2008, pp. 507–510. ACM Press (2008)
16. Harrison, C., Hudson, S.: Using shear as a supplemental two-dimensional input channel for rich touchscreen interaction. In: Proceedings of the CHI 2012, pp. 3149–3152. ACM Press (2012)
17. Heo, S., Lee, G.: Forcetap: extending the input vocabulary of mobile touch screens by adding tap gestures. In: Proceedings of the MobileHCI 2011, pp. 113–122. ACM Press (2011)
18. Heo, S., Lee, G.: Force gestures: augmenting touch screen gestures with normal and tangential forces. In: Proceedings of the UIST 2011, pp. 621–626. ACM Press (2011)
19. Heo, S., Lee, G.: ForceDrag: using pressure as a touch input modifier. In: Proceedings of the OzCHI 2012, pp. 204–207. ACM Press (2012)
20. Heo, S., Lee, G.: Indirect shear force estimation for multi-point shear force operations. In: Proceedings of the CHI 2013, pp. 281–284. ACM Press (2013)
21. Heo, S., Han, J., Lee, G.: Designing rich touch interaction through proximity and 2.5D force sensing touchpad. In: Proceedings of the OzCHI 2013, pp. 401–404. ACM Press (2013)

22. Herot, C., Weinzapfel, G.: One-point touch input of vector information for computer displays. In: Proceedings of the SIGGRAPH 1978, pp. 210–216. ACM Press (1978)
23. Hilliges, O., Izadi, S., Wilson, A.D., Hodges, S., Garcia-Mendoza, A., Butz, A.: Interactions in the air: adding further depth to interactive tabletops. In: Proceedings of the UIST 2009, pp. 139–148. ACM (2009)
24. Hinckley, K., Song, H.: Sensor synaesthesia: touch in motion, and motion in touch. In: Proceedings of the CHI 2011, pp. 801–810. ACM Press (2011)
25. Hirsch, M., Lanman, D., Holtzman, H., Raskar, R.: BiDi screen: a thin, depth-sensing LCD for 3D interaction using light fields. ACM Trans Graph 28(5), 159:1–159:9 (2009)
26. Leap Motion Controller, Leap Motion. https://www.leapmotion.com/product/
27. Lee, B., Lee, H., Lim, S.-C., Lee, H., Han, S., Park, J.: Evaluation of human tangential force input performance. In: Proceedings of the CHI 2012, pp. 3121–3130. ACM Press (2012)
28. Lee, G., Lee, S., Bang, W., Kim, Y.: A TV pointing device using LED directivity. In: IEEE International Conference on Consumer Electronics (ICCE 2011), pp. 619–620. IEEE Press (2011)
29. Logitech, Wireless Touchpa. http://www.logitech.com/en-us/product/touchpad-t650/
30. Marquardt, N., Jota, R., Greenberg, S., Jorge, J.A.: The continuous interaction space: interaction techniques unifying touch and gesture on and above a digital surface. In: Campos, P., Graham, N., Jorge, J., Nunes, N., Palanque, P., Winckler, M. (eds.) INTERACT 2011, Part III. LNCS, vol. 6948, pp. 461–476. Springer, Heidelberg (2011)
31. Microsoft Research, Project Gustav. http://research.microsoft.com/en-us/projects/gustav/
32. Minsky, M.: Manipulating simulated objects with real-world gestures using a force and position sensitive screen. In: Proceedings of the 11th Annual Conference on Computer Graphics and Interactive Techniques, pp. 195–203. ACM Press (1984)
33. Miyaki, T., Rekimoto, J.: GraspZoom: zooming and scrolling control model for single-handed mobile interaction. In: Proceedings of the MobileHCI 2009, p. 11. ACM Press (2009)
34. Ramos, G., Boulos, M., Balakrishnan, R.: Pressure widgets. In: Proceedings of the CHI 2004, pp. 487–494. ACM Press (2004)
35. Ramos, G., Balakrishnan, R.: Pressure marks. In: Proceedings of the CHI 2007, pp. 1375–1384. ACM Press (2007)
36. Rekimoto, J.: SmartSkin: an infrastructure for free hand manipulation on interactive surfaces. In: Proceedings of the CHI 2002, pp. 113–120. ACM Press (2002)
37. Rendl, C., Greindl, P., Probst, K., Behrens, M., Haller, M.: Presstures: exploring pressure-sensitive multi-touch gestures on trackpads. In: Proceedings of the CHI 2014, pp. 431–434. ACM Press (2014)
38. Research In Motion, BlackBerry Storm 2. http://worldwide.blackberry.com/blackberrystorm2/storm
39. Rosenberg, I., Perlin, K.: The UnMousePad: an interpolating multi-touch force-sensing input pad. ACM Trans. Graph. (TOG). 28(3), 65:1–65:9 (2009). ACM Press
40. Samsung, Galaxy S4. http://www.samsung.com/global/microsite/galaxys4/lifetask.html#page=airview/
41. Synaptics, ForcePad. http://www.synaptics.com/en/forcepad.php
42. Takeoka, Y., Miyaki, T., Rekimoto, J.: Z-touch: an infrastructure for 3d gesture interaction in the proximity of tabletop surfaces. In: Proceedings of the ITS 2010, pp. 91–94. ACM Press (2010)

43. Tsukada, Y., Hoshino, T.: Layered touch panel: the input device with two touch panel layers. In: Extended Abstracts, Proceedings of the CHI 2002, pp. 584–585. ACM Press (2002)
44. Wacom, Intuios Pro. http://www.wacom.com/en/us/creative/intuos-pro-m/
45. Westerman, W.: Hand Tracking, Finger Identification and Chordic Manipulation on a Multi-Touch Surface. Ph.D. thesis, University of Delaware (1999)
46. Wilson, A.D., Benko, H.: Combining multiple depth cameras and projectors for interactions on, above and between surfaces. In: Proceedings of the UIST 2010. ACM Press (2010)
47. Wu, M., Shen, C., Ryall, K., Forlines, C., Balakrishnan, R.: Gesture registration, relaxation, and reuse for multi-point direct-touch surfaces. In: Proceedings of the TableTop 2006, p. 8. IEEE Press (2006)
48. Yu, C., Tan, X., Shi, Y., Shi, Y.: Air finger: enabling multi-scale navigation by finger height above the surface. In: Proceedings of the UbiComp 2011, pp. 495–496. ACM Press (2011)
49. Zhai, S.: Human Performance in Six Degree of Freedom Input Control. Ph.D. thesis, University of Toronto (1995)

Video Gaming

Enhancing Spatial Perception and User Experience in Video Games with Volumetric Shadows

Tuukka M. Takala[1(✉)], Perttu Hämäläinen[1], Mikael Matveinen[1],
Taru Simonen[1], and Jari Takatalo[2]

[1] Department of Media Technology, Aalto University, Espoo, Finland
{tuukka.takala,perttu.hamalainen,mikael.matveinen,
taru.simonen}@aalto.fi
[2] POEM Research Group, University of Helsinki, Helsinki, Finland
jari.takatalo@helsinki.fi

Abstract. In this paper, we investigate the use of volumetric shadows for enhancing three-dimensional perception and action in third-person motion games. They offer an alternative to previously studied cues and visual guides. Our preliminary survey revealed that from the games that require Kinect, 37 % rely primarily on a third-person view and 9 % on a first-person view. We conducted a user study where 30 participants performed object reaching, interception, and aiming tasks in six different graphical modes of a video game that was controlled using a Kinect sensor and PlayStation Move controllers. The study results indicate that different volumetric shadow cues can affect both the user experience and the gameplay performance positively or negatively, depending on the lighting setup. Qualitative user experience analysis shows that playing was found to be most easy and fluent in a typical virtual reality setting with stereo rendering and flat surface shadows.

Keywords: Depth perception · Depth cues · Stereoscopy · Games · 3D user interface · Game experience · Volumetric shadows

1 Introduction

Precise 3D input devices have become mainstream thanks to low-cost game hardware like Microsoft Kinect and PlayStation Move. However, the dream of immersive first-person virtual reality (VR) has not come true in the household, because display technology is lagging behind. Meanwhile, commercial motion games use a single television with the game world typically portrayed in a first-person view (Rise of Nightmares), a third-person over the shoulder view (Kinect Star Wars, Fighters Uncaged, The Fight: Lights Out) or a third-person mirror-like view (Your Shape Fitness Evolved, Dance Central).

Our work is motivated by our experiences from prototyping a third-person Kinect action game where the object was to reach static and moving targets. In early tests using a television as the display, estimating distances and hitting targets in 3D seemed

© Springer International Publishing Switzerland 2015
T. Wyeld et al. (Eds.): OzCHI 2013, LNCS 8433, pp. 91–113, 2015.
DOI: 10.1007/978-3-319-16940-8_5

very difficult. It was clear that some form of additional cues or guidance was needed, preferably without the need for a stereo 3D display to reach a wider audience.

This paper presents a study that compares both proven and novel cues for enhancing target reaching and intercepting in a third-person perspective. Reaching and intercepting are tasks integral to most fighting, sports, and action games. Our main contribution is the study of volumetric shadows. We wanted to avoid possibly alienating and "brute force" visual guides such as rendering the predicted path of an object in the game world. We were particularly intrigued by the possibility of designing lighting so that it *enhances gameplay and creates positive user experience*.

According to our review of the literature, the use of volumetric shadows has not been studied in positioning or interception tasks.

2 Background

The human visual system combines a number of cues with information about the posture of the eyes to compute the distance of objects [1–3]. Cutting and Vishton [1] list a total of 15 sources of information about three-dimensional layout and distance: accommodation, aerial perspective, binocular disparity, convergence, height in visual field, motion perspective, occlusion, relative size, relative density, linear perspective, light and shading, texture gradients, kinetic depth, kinetic occlusion and disocclusion, and gravity. Increasing the amount of information generally increases the accuracy of depth judgments [1].

The effect of visual cues such as binocular disparity on motor behavior has been researched extensively. The importance of cues varies depending on the activity or sport. For example, Heinen and Vinken [4] report that binocular vision is not necessary for experienced gymnasts. On the other hand, Laby et al. [5] report better stereoacuity in Olympic-level soccer and softball athletes compared to archers, and various ball catching and hitting studies have found binocular vision superior to monocular [6–9]. However, monocular catching and hitting is still possible, and the importance of binocular vision varies as a function of, e.g., target size and speed [6, 8].

In the realm of computer graphics, Wanger et al. [3] found shadows and perspective to help in a positioning task conducted using a single 2D display. Their study required the participants to match the position, rotation and size of an object with a reference object. In the context of this paper, moving the avatar's hand to touch a target can also be regarded as a position matching task. Wanger's follow-up study [10] found shadow sharpness and size to have no significant effects on estimating object size and position. Hubona et al. [11] reported that in monocular viewing, shadows enhance positioning accuracy to a level equivalent to stereo 3D (S3D) viewing without shadows. However, they also found that in S3D shadows increased both accuracy and response times. Non-realistic depth cues have been explored by Glueck et al. [12], who introduced virtual position pegs standing on a multiscale reference grid.

There are many studies of using head-tracking-induced motion parallax to aid depth perception: Teather and Stuerzlinger [13], and Boritz and Booth [14] have investigated both S3D and head tracking in positioning tasks, and while their results support the importance of S3D, no significant effects were found for head tracking. However, it has also been shown that head tracking can help more than S3D in other tasks [15].

It should be noted that humans use focal and ambient vision differently for movement control. Focal vision is specialized for object identification and conscious movement control, and ambient vision is specialized for unconscious control based on, e.g., optical flow [16]. Conscious processing of focal vision is slow compared unconscious movement control based on ambient vision [17]. In commercial motion games and our tests using a single television about 2–3 m away from the user, the display is probably too small for ambient vision to be of much use.

The unnatural third-person perspective also changes the optical flow patterns. Although a first-person view could have less problems, being closer to our natural way of seeing the world, we agree with Oshita [18] that a third-person view is optimal for many full body action games, because it allows the players to fully see how their movements are mapped to the avatar's movements and how the avatar reacts to the environment.

We went through a list of 95 games that required Kinect [19] and viewed their gameplay videos online: 35 of the games relied primarily on a third-person view, 9 relied primarily on a first-person view, 41 offered a mirror-like view (usually the mirror image was very small), and the other 10 games involved primarily 2D pointer interaction or did not fit into our categorization. Existing games often make it easier to hit or intercept objects using graphical guides. Sport Champions: Table Tennis draws the predicted path of the ball as a 3D curve in the game world. The Kinect Sports goalkeeping game transforms an interception task into a reaching task by displaying a static marker that shows where the ball will be a moment later.

In the onset of our study we conducted preliminary tests with a third-person Kinect action game. Using a 2D television as the display, estimating distances and hitting targets in 3D seemed much more difficult than in real life or in an immersive VR system like the CAVE [20]. We found it possible to learn to hit static objects through trial and error, but moving objects were more often missed than hit. There was an apparent need for some form of additional cues or guidance, preferably without resorting to S3D displays so that wider audiences could be reached. This led us to experiment with volumetric shadows as depth cues. Our hypothesis is that volumetric lighting can be used to enhance spatial perception in real-time 3D applications. The purpose of our study was to test that hypothesis and to evaluate how different depth cues affect user experience.

2.1 Volumetric Shadows

In computer graphics, realism of rendered images can be improved by taking into account light scattering effects of air and other participating media [21]. This results as images where visible beams of light emanate from the light source. If there are objects blocking the light beams, then no light scattering will occur in the umbral regions behind the objects, and so called volumetric shadows (Fig. 1) will appear [22].

For the sake of clarity, we use the term *surface shadow* to describe ordinary shadows that occur on illuminated surfaces when direct illumination from a light source

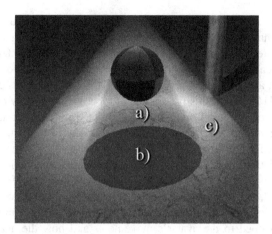

Fig. 1. (a) Volumetric shadow, (b) surface shadow, (c) volumetric light shaft.

is blocked by an object. It is worth noting that a light source can cast both surface shadows and volumetric shadows at the same time, as seen in Fig. 1. To distinguish from standard light sources that emit only direct light, we use the term *volumetric light source* for light sources that feature volumetric light shafts and shadows.

Surface shadows are prevalent in 3D video games, whereas few games use volumetric shadows. Those games that have volumetric shadows use them as mere eye-candy or for increased graphical immersion (e.g. Uncharted 3, F.E.A.R. 2). Alan Wake is a rare exception, as volumetric light sources play an important role in its gameplay, where it is necessary to illuminate the player character's adversaries in order to defeat them.

Previous lighting and shadow studies about depth perception have been mostly limited to surface shadows. A surface shadow can resolve the depth/size ambiguity, but it is not ideal if the object of interest is far from the surface, because (1) the shadow cannot be seen accurately while keeping ones gaze fixed on the target and (2) the shadow may end up occluded or outside the camera view. In contrast, a volumetric shadow traces the linear path between the object and the surface shadow that it casts. By doing so, the volumetric shadow also points into the direction of the light source.

As we stated, volumetric lighting and volumetric shadows are a result of light's interaction with participating media. There exists a multitude of research on participating media rendering techniques [23]. This research focuses almost exclusively on rendering quality and performance, and user studies about the effect of such advanced rendering techniques on depth perception are rare. However, a few such studies can be found in the field of medical imaging, where it can be a matter of life and death to understand 2D renditions of complicated 3D scans. Many rendering techniques for medical scans provide a translucent appearance that conveys depth by simultaneously revealing several tissue layers that otherwise would be occluded [24–26]. A related, albeit non-photorealistic technique for accentuating depth information is the use of volumetric halos [27, 28].

In our literature review of volumetric shadows, we came to the same conclusion as Lindemann and Ropinski, who stated that "with respect to volume rendering, perceptual studies are rather scarce" [29]. The existing few user studies that involve volumetric shadows always employ them as part of a more complex lighting scheme, and even then their purpose is limited to conveying depth information about translucent materials by increasing color dynamics of the rendered images [26, 29–32]. We found no studies where volumetric shadows would have been used for positioning or interception tasks. As far as we can tell, our study is the first one to examine volumetric light shafts as spatial cues for aiming and understanding relative positions between objects.

2.2 User Experience and Gameplay

Earlier video game lighting research has explored how lighting color affects gameplay performance and emotion [33], and how lighting can be automatically adjusted to accommodate dramatic, aesthetic, and communicative goals of a game [34]. Only a little is known how different depth cues affect user experience (UX) in games. Here, we refer the term UX as a subjective experience that stems from the use of technology that is in our case gameplay.

The concept of presence, namely the sense of being in a mediated environment, is broadly studied in different media [35]. Both S3D displays [36, 37] and head tracking [38] have been found to enhance presence. Spatial awareness, attention, and realness/ naturalness form the perceptual "Big three" sub-components of physical presence [39]. In addition to users' perceptions, their evaluations of the provided action are needed to study UX [37, 40].

The concept of flow provides a good framework to analyze human activity in many contexts. Csikszentmihalyi [41] defines flow as a positive and enjoyable experience stemming from an interesting activity that is considered worth doing for its own sake. In the core of the theory are the cognitively evaluated challenges provided by the activity and the skills possessed by the respondents. Flow evolves when a person evaluates both the challenges and the skills as being high and in balance. More specifically, the antecedents of flow – i.e. evaluated skills and challenges in the situation – indicate the quality of user experience [42]. For example, whether participants are experiencing mastery (skills above the challenges) or coping (skills below the challenges). In digital games, the mastery situation is found to have a positive effect on the enhanced motivation to continue on playing and to play again [43]. Thus, the flow framework integrates emotional and motivational layers on top of the perceptual-cognitive evaluation process.

3 User Study

We organized a user study in which participants played a 3D video game in six different experimental conditions. We advertised the study via mailing lists and social media, and recruited a total of 35 participants from our university; students, researchers, and staff. Each participant received a movie ticket voucher after taking part in the study.

3.1   Game

The game used in the study was a simplified, video game version of wall tennis implemented with the Unity game engine. The player's in-game avatar held a paddle racket in each hand, standing in front of an archery target with a diameter of 3 m. The rackets were represented and tracked with PlayStation Move controllers. The avatar was controlled via OpenNI library using a Kinect sensor.

The objective of the game was to strike virtual balls and hit as close to the archery target's center (bull's-eye) as possible. Hitting directly at the bull's-eye gave 100 points. The number of awarded points decreased linearly to zero on a disk with a radius of 4 m, so that points were given even when hitting outside the archery target. Hit score was displayed on screen for two seconds after a hit was registered, so that players would strive to get better scores.

Table 1. Depth cues among the study conditions.

Depth cue \ Condition	0	1	2	3	4	5
Stereo 3D		x				x
Head tracking		x	x	x	x	x
Surface shadows		x	(x)	(x)	x	(x)
Ball volumetric light				x		x
Racket volumetric lights			x			
Above volumetric light					x	

Our game had six different graphical modes that acted as test conditions in our study. Each condition had a unique set of depth cues, combining traditional cues and our volumetric shadows (Table 1). Screenshots of different conditions are presented in Fig. 2. Condition 0 acted as a baseline, as it had the least amount of depth cues, providing only perspective cues. Condition 1 represented the "industry standard" of VR graphics with surface shadows and S3D. In Condition 2 both rackets had an omnidirectional volumetric light source with a unique color. In Condition 3 each ball acted as an omnidirectional volumetric light source. Condition 4 presented a volumetric light source placed high above the playing area. Condition 5 had S3D; otherwise it was identical to Condition 3. Surface shadows in conditions 2, 3, and 5 were fainter than in conditions 1 and 4, due to the additional lights.

We made two design decisions for easing the learning curve of aiming and striking: (1) There was no gravity in the physics simulation and striking a ball sent it on a linear path with constant velocity; (2) Balls interacted only with the rackets and the in-game avatar passed through them without effect. Bouncing from ground and walls was left enabled for facilitating a basic sense of physical immersion.

The game presented only one ball at any given moment. After striking a ball and registering a hit or miss, the ball would disappear. Before a new ball would appear, the player had to return to the center of the playing area, marked by a circular object. For our study this meant that the event of striking a ball became a repeatable trial with

a) Condition 0.   b) Condition 1.

c) Condition 2.   d) Condition 3.

e) Condition 4.   f) Condition 5.

Fig. 2. Six graphical modes that were the test conditions of our user study.

always the same outset, regardless of the participant, previously hit balls, or other conditions. The game had both static balls hovering in the air and balls that appeared from either side of the archery target, moving linearly through the playing area. In the latter case, the movement was constant with the speed of 2 m/s, which was determined to be challenging enough in our initial testing phase.

Natural Aiming and Depth Cues. As seen from Fig. 2, volumetric lighting adds a sense of depth in conditions 2, 3, 4, and 5 of the game. Volumetric shadows in conditions 2, 3, and 5 form a cone frustum that is aligned along the line between the ball and the racket. In Condition 2, the light source in the racket casts a surface shadow from the ball that can be used as a laser sight to aim the ball. A high-above volumetric light source in Condition 4 creates volumetric shadows that bind objects to their surface shadows on the ground, providing a natural depth cue alternative to the non-realistic position pegs of Glueck et al. [12].

3.2   Equipment

The game ran 50 frames per second on a computer with Windows 7, Intel Core2 6600, 3 GB of RAM, and Nvidia GeForce 8800GTX. The game was displayed on a 55" Panasonic TX-L55ET5Y television that could output S3D through passive, circularly polarized stereo-glasses. PlayStation Move controllers were connected to our computer via Move.me software running on a PlayStation 3 that was equipped with a PlayStation Eye camera. The coordinate system between Move controllers and Kinect was calibrated with RUIS library [44].

3.3   Environment

Playing area in front of the TV was marked with a 2 m by 2 m floor mat, from which the game could be played without stepping outside. Illuminance within that area varied between 235–320 lux, as measured with Konica Minolta Chroma Meter CL-200. An illuminance between 95–135 lux was measured by the back wall that was seen as background by Kinect and PlayStation Eye camera. The playing area center was at distance of 3.2 m from the TV. There the luminous intensity was measured to be 157 cd/m2, by pointing Konica Minolta Luminance Meter LS-110 at a white TV screen. The luminous intensity through the passive stereo glasses was 65 cd/m2.

3.4   Design and Procedure

We used a randomized within-subject design with an incomplete counterbalancing, where the participants were exposed to each of the six conditions in a random order. Every other participant was exposed to the conditions in a reverse order compared to the previous participant, for counterbalancing learning effect over the conditions [45]. We paid special attention that the participants would not discover our hypothesis or other study details before or during the study; we did not mention lights, shadows, depth, or aiming cues in our interview questions. Instead we let the participants report their findings in their own words, discussing only topics that they had brought up themselves.

Metrics. The following information was collected from each of the participants: vision, background, and subjective UX evaluations after each test condition. Because of the large number of test conditions, we attempted to keep the test procedure simple, but to still get as rich UX descriptions from the participants as possible. Thus, we used both qualitative interviews and quantitative scales to assess perceptions and actions related to UX.

After each condition, participants evaluated their "overall feeling" (0–10 Likert), "challenges of the task" (1–7 Likert), and their own levels of skills (1–7 Likert). Our idea with overall feeling was to give the participants a chance to freely rate each condition. It provided a good one-dimensional comparison for the multidimensional interview data. In the past, we have used challenges and skills measures to empirically present Csikszentmihalyi's [41] four channel flow model [46]. We divided 2,182 participants based on their challenge and skill/competence evaluations and showed how participant's cognitive evaluations affected their emotional outcomes in digital games. The flow-space thus formed enables us to evaluate whether participants are experiencing mastery or coping while conducting the experiment.

We have tested participants' challenge and competence evaluations in experiments conducted in between-subjects design. Our results indicate, that if different groups of participants are engaged with the same digital game [47] or a task in a virtual environment (VE) [48], then they tend to evaluate the challenges of the situation similarly. However, participants playing the digital game in the laboratory evaluated their level of competence lower compared to those playing the same game at home. Gamers' background data revealed that at home the gamers were more experienced with the game. This obvious finding validates the used flow-space measures.

In a VE, participants moving fluently evaluate their level on competence higher compare to those being stationary (that is, moving less fluently). In this case, the background data did not provide any explanation for the finding. It seems that some participants learned to utilize the novel interaction device in VE faster than the others. Both the subjective competence evaluation and objective movement data revealed the difference among the participants and again validated the flow-space measures.

In addition, flow-space provides valuable information about the experiential process. We have demonstrated how the UX in the same digital game evolves during the first hour of play [49]. Supplementing competence and challenge measures with the gamer background, game performance and subjective interviews provided us a detailed description of how the other gamer was ready to quit and the other was just warming up. Although an individual level inspection often turns into a qualitative analysis, this example further substantiates the added value of the flow-space in evaluating human-technology interaction and games.

The participants were interviewed and asked to describe the environment and the task. Different descriptions were compressed into 17 dichotomous variables. Each of the 17 variables had at least ten mentions among the participants. The 17 variables thus formed were analyzed in a correspondence analysis (CA), which is a multivariate descriptive data analytical technique for categorical data. CA shares similarities with the principal components analysis, which applies to continuous data. CA is helpful in depicting the relationship between two or more categorical variables in a 2-dimensional chart, summarizing and illustrating similarities and differences between categories and the associations between them [50].

Finally, the participants rated their feelings of Pleasure/Valence, Arousal, and Dominance/Control (PAD) using Self-Assessment Manikin's (SAM) [51]. The PAD profile describes participants' degree of valence, level of activeness of an emotion, and experienced sense of control. Flow theory [41] acknowledges PAD emotions as important flow factors.

Procedure. First the participants with corrected vision were asked to wear their eye-wear if that helped them to see the TV better. Next we administered the standard TNO stereoscopic vision test [52], in order to determine how sensitive the participants' stereoscopic vision was on a scale of 15–480 arc seconds.

Before starting, participants practiced the game for 5 min in a graphical setting that had only surface shadows and monocular rendering. They were instructed that the goal of the game was to strike the ball so that it would hit as close to the bull's-eye as possible without bouncing it off from the ground or walls. The participants were also informed about tracking problems that would occur if the PlayStation Move controllers were turned away from PlayStation Eye camera's view. This was a problem for those participants who were using the controllers like a tennis racket, swinging their hand backwards and swiveling the controller away from the PlayStation Eye camera before striking. The problem was pronounced when striking balls in the farther edges of the playing area.

After practice, the participants were exposed to six different conditions, each consisting of 30 trials of striking the game's virtual ball. Half of the balls were static and half were moving linearly, and these two sets were constant for each participant

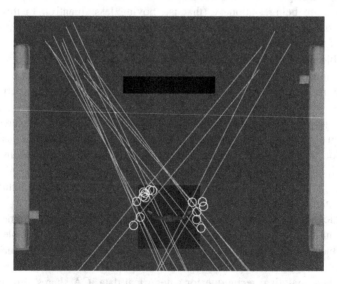

Fig. 3. Static ball positions are displayed with white circles, and moving ball trajectories with blue lines. The red square represents the playing area (Colour figure online).

and each condition, but the order of balls was randomized. Figure 3 illustrates the position of the static balls and trajectories of the moving balls. The static and moving balls had a varied distance from the ground and the moving balls' trajectories were parallel to ground plane.

Striking through 30 balls in each condition took on average 4 min. Participants wore the stereo-glasses even during the conditions that did not have S3D. After each condition, we interviewed the participant about the experience. These interviews lasted on average 8 min each. The mean duration for the whole experiment was 90 min.

4   Results and Analysis

When calculating study results, we included 30 people with stereoscopic acuity better or equal to 120 arc seconds from the total of 35 participants. This way the possible differences between S3D and monocular rendering would not be obscured by participants with poor stereoscopic vision.

4.1   Participant Demographics

Most of the 30 participants (90 %) were Finnish speaking students or research staff of the Aalto University. The mean age of the participants was 27.7 years (SD = 4.04 years). Participants were rather familiar with the commercially available 3D user interfaces (3DUI) for digital games: 27 % had used Sony Move, 43 % Microsoft Kinect, and 87 % Nintendo Wii. However, 23 % reported to have no experience on 3DUIs and most of them evaluated their level of experience either basic (63 %) or intermediate (14 %). No one reported to be an expert in 3DUIs.

Participants reported to play digital (i.e., computer or video) games with varying intensity: 10 % played at least once every two days, 27 % less than that but at least once a month, and 47 % infrequently. Only one participant reported playing daily and four reported that they never play digital games.

Regarding actual physical sports similar to our test set-up, majority of the participants had at least tried badminton (94 %), tennis (83 %), table tennis (91 %), and volleyball (74 %). Approximately 89 % of the participants were right-handed. Although 69 % of participants reported doing some sport often (but less than 50 % of days), only one of them was a frequent badminton player. All in all seven participants reported to play badminton at some frequency, which was the most popular sport among the participants with similarities to our game. Thus, our sample represented well the casual users of 3DUIs and did not include an over-representation of either hard-core gamers or hard-core racket/ball game players.

Participants included in the study consisted of 24 males and 6 females, and hence our results are more representative of the male gender. However, there were no significant differences between genders with regards to our UX metrics (Likert ratings and interviews).

4.2   Gameplay Performance Results

We explored statistically significant differences between conditions with the Kruskall-Wallis test, Friedman test, and a 3-way ANOVA whose factors were condition index, ball index, and subject index. Post hoc tests were applied with a p-value of 0.05 using Tukey-Kramer correction. Results were obtained with Matlab's Statistics Toolbox.

Aiming Accuracy. We measured balls' hit accuracy with two variables: (1) Distance of the hit location on the archery target plane from the bull's-eye. If a ball had missed the target plane or hit the ground or side walls, its distance was set as the maximum distance that was recorded. By doing this, even the balls that missed the target plane were taken into account when examining hit accuracy between conditions with a Kruskall-Wallis test. (2) We used a 3-way ANOVA over all the balls where binary outcome of the hit event was the dependent variable. Static and moving balls were examined separately.

Figure 4a is a boxplot presenting the number of static balls that hit the archery target without bouncing from ground or walls. No significant differences were found with the 3-way ANOVA, Kruskall-Wallis, or Friedman tests.

Number of moving balls that hit the archery target can be seen in the boxplot chart of Fig. 4b. By comparing Figs. 4a and b it is clear that the participants were able to hit the target much more often when striking static balls. In fact, the participants missed many of the moving balls completely when trying to strike them (failed interception).

Significant main effect of conditions was found with Kruskall-Wallis in moving balls' hit distance from the bull's-eye ($\chi 2(5) = 17.7$, p = 0.003). The post hoc test showed that accuracy was worse in Condition 0 when compared both to conditions 1 and 4. These results were repeated by the 3-way ANOVA followed by a post-hoc test.

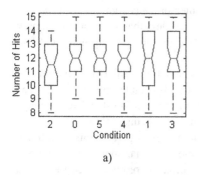
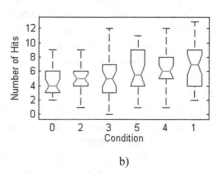

a)   b)

Fig. 4. Number of (a) static and (b) moving balls that hit the archery target, struck by 30 participants. The conditions are sorted by medians.

Interception Rate of Moving Balls. We measured interception rate with the number of moving balls intercepted by each participant under different conditions (Fig. 5). Interception rate was worst in Condition 0 as expected, followed by conditions 3, 5, and 2. Interestingly, interception rate was the highest in Condition 4, despite its monocular rendering. Condition 1 came second with its S3D.

Significant main effect of conditions was found with Friedman test ($\chi2(5) = 18.9$, p = 0.002). Post hoc test revealed that interception rate was higher in Condition 4 when compared both to conditions 0 and 3. No other statistically significant differences were found. These results were repeated when treating binary outcome of a moving ball's interception event as a dependent variable and applying the 3-way ANOVA over all the moving balls, followed by a post hoc test.

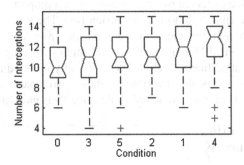

Fig. 5. Number of moving balls intercepted by each of the 30 participants. The conditions are sorted by medians.

Acquisition Time for Static Balls. We also examined static balls' acquisition times; i.e. the elapsed time from each ball's appearance till it was struck by the participant. We found significant main effect of conditions ($F(5, 2651) = 19.9, p < 0.001$) using the 3-way ANOVA. According to a post hoc test, acquisition time of Condition 2 was the longest (median of 3.1 s) of all the conditions (medians between 2.5 and 2.6 s). No other significant differences were found.

Kruskall-Wallis test revealed that those who reportedly used lighting cues for aiming (4 participants) had significantly better aiming accuracy ($\chi2(1) = 4.9, p = 0.027$) in Condition 2 than those who did not use lighting cues (26 participants). The mean test order number of Condition 2 was 3.3 for the former group and 4.0 for the latter, and thus the group using lighting cues had had a little more practice with the task. This fact and the small group size means that we cannot say conclusively if using shadows for aiming improved accuracy in Condition 2.

4.3   UX Results

Based on our interviews, four participants out of 30 discovered and used the racket light sources' shadows for aiming in Condition 2. Two additional participants made the discovery, but they did not continue to aim with the shadows as they did not find them beneficial. Four other participants mentioned that the rackets' volumetric light sources improved spatial perception. In Condition 3 the balls acted as volumetric light sources, and 13 participants reported that this enhanced spatial perception. Condition 4's high-above volumetric light source was mentioned by 12 participants to improve spatial perception.

UX Interviews. We calculated all UX results with SPSS, a software for statistical analysis. The 17 dichotomous variables representing participants' perceptions about the environment were analyzed in a correspondence analysis (CA). CA provides orthogonal dimensions that are extracted in order to maximize the distance between row and column points [50]. We used participants' perceptions as column variables, and the six lighting conditions as row variables.

Figure 6 presents the correspondence between the 17 perceptions and the six lighting conditions. We extracted two dimensions, with a proportion of 40 % of inertia to the dimension one and 28 % to the dimension two. The $\chi^2[96] = 286,4$, p < .001 supported the meaningful relationship between the row and column variables. Dimension one was named Diminished Sense of Depth - Exciting but Unnatural, and dimension two was called Interactively Fluent - Visually Distracting, according to the corresponding variables. As the descriptions show, participants integrate their perceptions and actions when they are asked to describe their interactive environment.

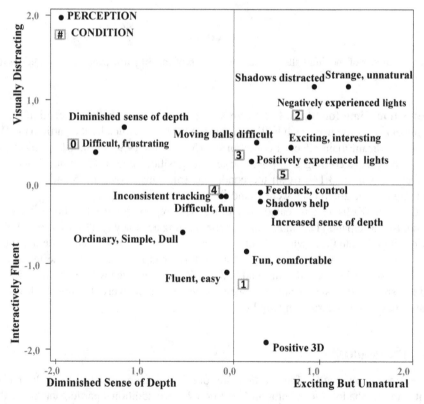

Fig. 6. Correspondence analysis showing the relationships between the 17 perception variables and the six conditions.

The inspection of the Fig. 6 shows that Condition 0 was perceived as spatially poor and difficult to play, mostly because of the lack of shadows. Condition 1 provided a positive 3D environment that was simple and sometimes even dull, in which playing was fluent and easy. Condition 2 was somewhat interesting and exciting, but for most of the participants the rackets' volumetric light sources and their colors were distracting and unnatural. This and the somewhat confusing shadows in Condition 2 made playing less comfortable and more difficult.

The rest of the conditions were closer to each other experientially. Although some of the participants in Condition 3 perceived the ball's volumetric light source and its colors positively, in most cases it diminished the spatial perception of the ball. This made hitting more difficult, especially with moving balls. On the contrary, volumetric light source in Condition 4 made hitting a bit fluent and easier, but still difficult. Some participants complained Condition 4 to be too simple and even dull. Because of the S3D, high spatial perception was mentioned more often in Condition 5 compared to Condition 3. However, the volumetric light in the ball divided participants' opinions: some of them liked it but some thought that it distracted them. Problems with racket tracking were reported quite equally across all the conditions, while Condition 1 received the largest number of mentions. Tracking problems may have been ignored more in other conditions because of the other, possibly stronger perceptions that caught the participants' attention.

UX Scales. First, we studied UX in the playing order of the conditions with General Linear Model Repeated Measures Anova (PASW Statistics 18, pairwise comparisons between conditions were applied with a p-value of 0.05 using Bonferroni correction). Overall feeling increased significantly as the experiment proceeded (Wilk's Lambda = .33, $F(5,25) = 10.10$, $p < .001$, $\eta2 = .67$). Similarly, the evaluated skills increased (Wilk's Lambda = .49, $F(5,25) = 5.17$, $p < .01$, $\eta2 = .51$). Since the participants evaluated challenges the same way throughout the test, they experienced a clear learning curve from coping towards mastery (Wilk's Lambda = .45, $F(5,25) = 6.01$, $p < .01$, $\eta2 = .55$). The sense of control was the only PAD scale that increased significantly when the experiment proceeded (Wilk's Lambda = .59, $F(5,25) = 3.52$, $p < .05$, $\eta2 = .41$). Notably, good overall feeling, skills, and sense of control were easier to obtain in some of the conditions than others, regardless of the order of the condition.

The six conditions were significantly different in overall feeling (Wilk's Lambda = .31, $F(5,25) = 11.10$, $p < .001$, $\eta2 = .69$). Pairwise comparison shows that the overall feeling was significantly the lowest in Condition 0. Overall feeling in Condition 1 was higher compared to conditions 0, 2, and 4. Conditions 2, 3, 4, and 5 scored equally.

The flow-space (Fig. 7) shows that the participants were coping in Condition 0, which was rated as significantly more challenging than all the other conditions but Condition 2. Conditions 1, 2, 3, 4, and 5 did not differ from each other in this regard (Wilk's Lambda = .54, $F(5,25) = 4.30$, $p < .01$, $\eta2 = .46$). Moreover, in Condition 1 the participants experienced mastery. The evaluated skills in Condition 1 were significantly higher than in conditions 0 or 2 (Wilk's Lambda = .61, $F(5,25) = 3.19$, $p < .05$, $\eta2 = .39$).

PAD profiles show that participants were equally aroused across conditions (Fig. 8). Condition 1 was the highest in valence and conditions 0 and 2 were the lowest (Wilk's Lambda = .24, $F(5,25) = 15.73$, $p < .001$, $\eta2 = .76$). There was no difference between conditions 3, 4, and 5 in valence. Condition 1 was also significantly higher in the sense of control compared to Condition 0 (Wilk's Lambda = .60, $F(5,25) = 3.29$,

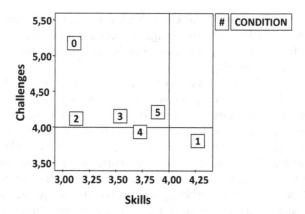

Fig. 7. The means of the skills and challenges of each condition plotted in the flow-space.

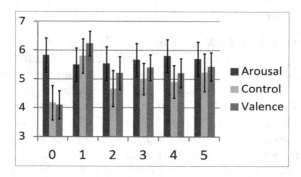

Fig. 8. The means and 95 % confidence intervals of the PAD profiles in the six conditions.

$p < .05$, $\eta2 = .40$). The three dimensions of the PAD profile reveal how equal degree of arousal across the conditions affects UX differently if it is accompanied either with low control and valence (Condition 0) or high control and valance (Condition 1).

Interview-Scale Correlations. Finally, we correlated participants' subjective descriptions and quantitative UX-scales across the six experimental conditions (Table 2). The descriptions were in line with the UX-scales deepening their information. For example, we can see that sense of depth, fluency, difficulty, sense of 3D, and lights, colors, and shadows help to constitute the generic "General Feeling" measure that is aligned with Valence. Furthermore, "Ordinary, simple, dull", "Strange, unnatural", Exciting, interesting", and both distracting lights and shadows provided descriptions that no scale could map. The correlations also show that the Challenge scale measured the difficulty of the task. Although Arousal scale did not correlate with any of the descriptions, integrating both the flow-space and the PAD profile data shows that low skills combined with high challenges and arousal led to a frustrating and unpleasant experience (Condition 0).

Table 2. Pearson correlations between the subjective descriptions and quantitative scales across the six experimental conditions (N = 180). Significant correlations were found with a 2-tailed test at 0.05 level (*) and 0.01 level (**).

	General Feeling	Valence	Arousal	Control	Challenge	Competence
Diminished sense of depth	−.291**	−.208**	.034	−.302**	.305**	−.181*
Fluent, easy	.237**	.241**	.066	.224**	−.203**	.059
Fun, comfortable	.125	.176*	−.043	.057	−.155*	.094
Ordinary, Simple, Dull	−.041	.013	.025	.131	.062	.038
Strange, unnatural	−.116	.000	.116	.030	.026	−.074
Inconsistent tracking	−.139	−.156*	−.046	−.046	−.018	−.074
Difficult, fun	.141	.162*	−.101	−.003	−.060	.057
Difficult, frustrating	−.359**	−.420**	−.072	−.311**	.222**	−.178*
Feedback, control	.116	.097	.053	−.035	−.168*	.083
Increased sense of depth	.247**	.262**	.013	.108	−.169*	.121
Positive 3D	.230**	.197**	−.029	.295**	−.120	.188*
Exciting, interesting	−.039	.073	−.058	−.009	−.017	−.046
Positively experienced lights & colors	.273**	.260**	−.008	.209**	−.159*	.115
Negatively experienced lights & colors	−.067	−.014	−.033	.057	.067	−.033
Moving balls difficult	.002	−.048	−.032	−.055	.060	.186*
Shadows help	.211**	.243**	.005	.162*	−.090	.068
Shadows distracted	−.013	−.020	−.058	−.075	.106	.023

Taking all the qualitative and quantitative UX measures into account reveals how each condition was perceived, evaluated, and finally experienced. The UX and its causes are rather univocal in conditions 0, 1, and 2. Condition 4 has some clear characteristics of its own, but conditions 3 and 5 are difficult to distinguish. These subjective findings are in line with our gameplay performance results.

5 Discussion

Compiling the UX and gameplay performance data that we gathered gives a rich description of the UX and gameplay in our six different conditions. Analysis of the interviews revealed both perceptual and action dimensions. Moreover, the analysis revealed the "big three" physical presence dimensions, that is spatial awareness (diminished sense of depth), attention (visually distracting), and realness/naturalness (exciting but unnatural) [39].

Condition 0 lacked shadows and was the worst condition in terms of UX and gameplay performance results. In Condition 2 some participants used shadows to aim the balls towards the bull's-eye, which contributed to the static ball acquisition time that was significantly the longest. Our results were inconclusive whether this aiming improved hit accuracy. Similar to Condition 0, the challenges and skills balance in Condition 2 was towards coping. Although the ratio between arousal and control was better balanced, playing was evaluated as uncomfortable and uneasy.

Participants achieved the best UX in Condition 1 with its typical VR setting of S3D and surface shadows; they experienced mastery (skills above the challenges) in the perceptually positive 3D environment. Condition 4 had volumetric shadows instead of S3D and it was the closest to Condition 1 in mastery and similar in other aspects: both received mentions about being fluent and easy to play, while being simple or even dull. On the other hand, conditions 1 and 4 had very different PAD profiles: equally high level of arousal was associated with lower valence and control in Condition 4, whereas in Condition 1 it was associated with higher control and valence. This difference might be related to the lower overall feeling in Condition 4.

Together conditions 1 and 4 were better or on par with the other conditions in gameplay performance results. No statistically significant differences were found between the two. Interestingly, Condition 4 had a significantly better interception rate compared to conditions 0 and 3, whereas the S3D-equipped Condition 1 did not. Since we did not have a test condition with monocular rendering and mere surface shadows from a high-above light source, it is not clear how much Condition 4's volumetric shadows might have improved the result against mere surface shadows.

S3D was the only setup distinction between conditions 3 and 5. Usually S3D increases the experience of physical presence [36, 37], but we did not find any clear differences in either the UX or gameplay performance results. This implies that the S3D's positive effect on depth perception in Condition 5 could have been diminished due to its volumetric lighting setup. Although the volumetric light sources used in conditions 3 and 5 were experienced mostly positively, it seems that they decreased the object-background contrast, thus diminishing the amount of depth cues and hindering spatial perception, which is possible according to a study by Schor and Wood [53]. We suspect that this is why the interception rate in Condition 3 was significantly lower than in Condition 4.

Conditions 1 and 5 were the only ones with S3D. There were no significant differences between these two conditions in quantitative UX or gameplay performance results. Mastery was experienced only in Condition 1 which received the most mentions about being fluent and easy. Condition 1 also had a unique PAD profile while Condition 5's profile resembled that of conditions 2, 3, and 4.

These pairwise comparisons between conditions 1 and 5 and conditions 3 and 4 suggest the following: volumetric light source inside the target object could (1) negatively affect the UX and (2) possibly impair gameplay performance when compared to a high-above volumetric light source. Our results imply that the high-above volumetric light source is the best choice from the three different types of volumetric lighting setups of our game in terms of UX and gameplay performance results.

We found no significant differences between conditions when examining hit accuracy of static balls. It appears that the dominant depth cue with static balls was occlusion; participants often moved their hand in xy-plane until it was in front or behind of the ball and then adjusted the z-position until the ball was hit.

Our task of striking balls towards a bull's-eye might not be optimal for eliciting depth perception related performance differences, although many motion games often have a simplified version of this task. Precise aiming in our game required the racket to be swung in 3D so that its collision with the ball would result in a trajectory towards the bull's-eye. For this the participants had to sense the 3D location of the ball and the 6D pose of the racket simultaneously. We suspect that the positioning and scaling tasks from prior studies [2, 3] could have led to more clear results.

Color of lighting was notably different in conditions 2, 3, and 5. Due to the high number of conditions and already long experiment duration, we decided to focus on depth cues and ignored colors as possibly contributing factors.

5.1 Lighting Guidelines for Improving Spatial Perception

Based on our results and observations during the study, we composed a short list of guidelines to aid lighting design in 3D applications where spatial perception is important:

1. Objects should be well contrasted against their background. Volumetric lighting and other lighting techniques can reduce this contrast and make it difficult to clearly distinguish visual border of an object (conditions 3 and 5).
2. Surface shadows that are meant to improve depth perception should be clearly visible. Additional illumination such as that of volumetric lighting can weaken these shadows (conditions 2, 3, and 5).
3. Two moving light sources can distract the user and negatively affect the UX (Condition 2). This is in line with a study by Hubona et al. [2], who reported that two light sources can impair task performance.

6 Conclusion

In this paper, we introduced novel lighting cues that can be used to assist reaching, interception, and aiming tasks, as well as enhance spatial perception. The cues are natural and blend into the rendered images because they are based on realistic rendering of volumetric lighting. This offers an alternative to traditional visual guides that are augmented over images and may appear out of place.

We presented a user study with 30 participants where the lighting cues were tested. Our results indicate that volumetric shadows can affect gameplay performance and UX positively or negatively, depending on the lighting setup. Statistically significant differences in our gameplay performance results imply that volumetric shadows can affect depth perception. A high-above volumetric light source with monocular rendering (Condition 4) did not differ from our best S3D setup with mere surface shadows (Condition 1) in terms of gameplay performance. Conversely, Condition 4 had a significantly better interception rate when compared to two other conditions whereas Condition 1 did not. Further studies are needed to quantify how much volumetric shadows can increase depth perception in monocular and S3D conditions when compared to surface shadows.

We analyzed UX with CA, Likert-scales, and flow-space metrics. Nearly half of the participants reported enhanced spatial perception in conditions with volumetric lighting. The use of volumetric light sources in our game divided the study participants' experiences however: some were pleased with exciting and interesting lighting conditions while others were distracted by them. We found indications that a poor choice of volumetric lighting could diminish S3D's positive effects on UX and depth perception. Overall, the most pleasing game experience was achieved with S3D and surface shadows (Condition 1).

Our study sets a starting point for further research on volumetric shadows as visual guides. Future studies need to confirm our findings for applications with first-person viewpoints. Moreover, future work should explore different aspects of UX. For instance, the concept of self-presence and its three subcomponents, namely proto (body-schema), core (emotion-driven) and extended (identity-relevant) [54] provide a noteworthy addition to physical presence and cognitive-emotional flow measures used in this study.

Acknowledgments. This work was supported by Finnish Doctoral Program in User-Centered Information Technology (UCIT) and Helsinki Institute of Science and Technology Studies (HIST).

References

1. Cutting, J.E., Vishton, P.M.: Perceiving layout and knowing distances: the integration, relative potency, and contextual use of different information about depth. Percept. space motion **5**, 69–117 (1995)
2. Hubona, G.S., Wheeler, P.N., Shirah, G.W., Brandt, M.: The relative contributions of stereo, lighting, and background scenes in promoting 3D depth visualization. ACM Trans. Comput. Hum. Interact. **6**, 214–242 (1999)
3. Wanger, L.R., Ferwerda, J.A., Greenberg, D.P.: Perceiving spatial relationships in computer-generated images. IEEE Comput. Graph. Appl. **12**, 44–58 (1992)
4. Heinen, T., Vinken, P.M.: Monocular and binocular vision in the performance of a complex skill. J. Sports Sci. Med. **10**, 520–527 (2011)
5. Laby, D.M., Kirschen, D.G., Pantall, P.: The visual function of olympic-level athletes—an initial report. Eye Contact Lens Sci. Clin. Pract. **37**, 116–122 (2011)
6. Bennett, S., van der Kamp, J., Savelsbergh, G.J.P., Davids, K.: Discriminating the role of binocular information in the timing of a one-handed catch. Exp. Brain Res. **135**, 341–347 (2000)
7. Van Hof, P., van der Kamp, J., Savelsbergh, G.J.P.: Three- to eight-month-old infants' catching under monocular and binocular vision. Hum. Mov. Sci. **25**, 18–36 (2006)
8. Mazyn, L., Lenoir, M., Montagne, G., Delaey, C., Savelsbergh, G.: Stereo vision enhances the learning of a catching skill. Exp. Brain Res. **179**, 723–726 (2007)
9. Bulson, R., Ciuffreda, K.J., Ludlam, D.P.: Effect of binocular vs. monocular viewing on golf putting accuracy. J. Behav. Optom. **20**, 31–34 (2009)
10. Wanger, L.R.: The effect of shadow quality on the perception of spatial relationships in computer generated imagery. In: Proceedings of 1992 Symposium on Interactive 3D Graphics, pp. 39–42 (1992)

11. Hubona, G.S., Shirah, G.W., Jennings, D.K.: The effects of cast shadows and stereopsis on performing computer-generated spatial tasks. IEEE Trans. Syst. Man Cybern. Part A Syst. Hum. **34**, 483–493 (2004)
12. Glueck, M., Crane, K., Anderson, S., Rutnik, A., Khan, A.: Multiscale 3D reference visualization. In: Proceedings of the 2009 Symposium on Interactive 3D Graphics and Games, pp. 225–232 (2009)
13. Teather, R.J., Stuerzlinger, W.: Guidelines for 3D positioning techniques. In: Proceedings of the Conference on Future Play, pp. 61–68. ACM, New York (2007)
14. Boritz, J., Booth, K.S.: A study of interactive 3D point location in a computer simulated virtual environment. In: Proceedings of the ACM Symposium on Virtual Reality Software and Technology, pp. 181–187. ACM, New York (1997)
15. Arthur, K.W., Booth, K.S., Ware, C.: Evaluating 3D task performance for fish tank virtual worlds. ACM Trans. Inf. Syst. **11**, 239–265 (1993)
16. Schmidt, R.A., Wrisberg, C.N.: Motor Learning and Performance. Human Kinetics Publishers, Champaign (2004)
17. Schneider, W., Shiffrin, R.M.: Controlled and automatic human information processing: I. Detection, search, and attention. Psychol. Rev. **84**, 1 (1977)
18. Oshita, M.: Motion-capture-based avatar control framework in third-person view virtual environments. In: Proceedings of the ACM SIGCHI International Conference on Advances in Computer Entertainment Technology, p. 2. ACM (2006)
19. Wikipedia: List of Kinect games. http://en.wikipedia.org/wiki/Kinect_games
20. Cruz-Neira, C., Sandin, D.J., DeFanti, T.A.: Surround-screen projection-based virtual reality: the design and implementation of the CAVE. In: Proceedings of the 20th Annual Conference on Computer Graphics and Interactive Techniques, pp. 135–142 (1993)
21. Nishita, T., Miyawaki, Y., Nakamae, E.: A shading model for atmospheric scattering considering luminous intensity distribution of light sources. In: ACM SIGGRAPH Computer Graphics, pp. 303–310 (1987)
22. Wyman, C., Ramsey, S.: Interactive volumetric shadows in participating media with single-scattering. In: IEEE Symposium on Interactive Ray Tracing, pp. 87–92 (2008)
23. Cerezo, E., Pérez, F., Pueyo, X., Seron, F.J., Sillion, F.X.: A survey on participating media rendering techniques. Vis. Comput. **21**, 303–328 (2005)
24. Ament, M., Sadlo, F., Weiskopf, D.: Ambient volume scattering. IEEE Trans. Vis. Comput. Graph. **19**, 2936–2945 (2013)
25. Yang, F., Li, Q., Xiang, D., Cao, Y., Tian, J.: A versatile optical model for hybrid rendering of volume data. IEEE Trans. Vis. Comput. Graph. **18**, 925–937 (2012)
26. Ropinski, T., Doring, C., Rezk-Salama, C.: Interactive volumetric lighting simulating scattering and shadowing. In: IEEE Pacific Visualization Symposium, pp. 169–176 (2010)
27. Bruckner, S., Groller, M.E.: Enhancing depth-perception with flexible volumetric halos. IEEE Trans. Vis. Comput. Graph. **13**, 1344–1351 (2007)
28. Tao, Y., Lin, H., Dong, F., Clapworthy, G.: Opacity volume based halo generation for enhancing depth perception. In: 2011 12th International Conference on Computer-Aided Design and Computer Graphics (CAD/Graphics), pp. 418–422. IEEE (2011)
29. Lindemann, F., Ropinski, T.: About the influence of illumination models on image comprehension in direct volume rendering. IEEE Trans. Vis. Comput. Graph. **17**, 1922–1931 (2011)
30. Boucheny, C., Bonneau, G.-P., Droulez, J., Thibault, G., Ploix, S.: A perceptive evaluation of volume rendering techniques. ACM Trans. Appl. Percept. (TAP) **5**, 23 (2009)
31. Šoltészová, V., Patel, D., Viola, I.: Chromatic shadows for improved perception. In: Proceedings of the ACM SIGGRAPH/Eurographics Symposium on Non-Photorealistic Animation and Rendering, pp. 105–116. ACM (2011)

32. Wang, L., Kaufman, A.E.: Lighting system for visual perception enhancement in volume rendering. IEEE Trans. Vis. Comput. Graph. **19**, 67–80 (2013)
33. Knez, I., Niedenthal, S.: Lighting in digital game worlds: effects on affect and play performance. CyberPsychology Behav. **11**, 129–137 (2008)
34. El-Nasr, M.S., Horswill, I.: Automating lighting design for interactive entertainment. Comput. Entertain. **2**, 15 (2004)
35. Lombard, M., Jones, M.T.: Identifying the (tele)presence literature. PsychNology J. **5**, 197–206 (2007)
36. IJsselsteijn, W., de Ridder, H., Freeman, J., Avons, S.E., Bouwhuis, D.: Effects of stereoscopic presentation, image motion, and screen size on subjective and objective corroborative measures of presence. Presence Teleoperators Virtual Environ. **10**, 298–311 (2001)
37. Takatalo, J., Kawai, T., Kaistinen, J., Nyman, G., Häkkinen, J.: User experience in 3D stereoscopic games. Media Psychol. **14**, 387–414 (2011)
38. Snow, M.P., Williges, R.C.: Empirical Models based on free-modulus magnitude estimation of perceived presence in virtual environments. Hum. Factors J. Hum. Factors Ergon. Soc. **40**, 386–402 (1998)
39. International Society for Presence Research: The Concept of Presence: Explication Statement. http://ispr.info/
40. Takatalo, J., Nyman, G., Laaksonen, L.: Components of human experience in virtual environments. Comput. Hum. Behav. **24**, 1–15 (2008)
41. Csikszentmihalyi, M.: Beyond Boredom and Anxiety. Jossey-Bass Publishers, San Francisco (1975)
42. Takatalo, J., Häkkinen, J.: Profiling user experience in digital games with the flow model. In: Proceedings of the Nordic Conference on Human-Computer Interaction (NordiCHI 14), Helsinki, Finland, pp. 26–30 (2014)
43. Ryan, R., Rigby, C., Przybylski, A.: The motivational pull of video games: a self-determination theory approach. Motiv. Emot. **30**, 344–360 (2006)
44. Takala, T.M., Pugliese, R., Rauhamaa, P., Takala, T.: Reality-based user interface system (RUIS). In: Proceedings of the IEEE Symposium on 3D User Interfaces 2011, pp. 141–142 (2011)
45. Dey, A.: Incomplete Block Designs. World Scientific Publishing, Singapore (2010)
46. Takatalo, J., et al.: Psychologically-based and content-oriented experience in entertainment virtual environments (2011)
47. Takatalo, J., Häkkinen, J., Kaistinen, J., Nyman, G.: User experience in digital games: differences between laboratory and home. Simul. Gaming **42**, 656–673 (2010)
48. Särkelä, H., Takatalo, J., May, P., Laakso, M., Nyman, G.: The movement patterns and the experiential components of virtual environments. Int. J. Hum Comput Stud. **67**, 787–799 (2009)
49. Takatalo, J., Häkkinen, J., Kaistinen, J., Nyman, G.: Presence, involvement, and flow in digital games. In: Bernhaupt, R. (ed.) Evaluating User Experience in Games, pp. 23–46. Springer, London (2010)
50. Greenacre, M.J.: Theory and applications of correspondence analysis. Academic press, London (1984)
51. Lang, P.J.: Behavioral treatment and bio-behavioral assessment: Computer applications. In: Sidowski, J.B., Johnson, J.H., Williams, T.A. (eds.) Technology in Mental Health Care Delivery Systems, pp. 119–137. Ablex Publishing Corporation, Norwood (1980)
52. Laméris Ootech: TNO Test for Stereoscopic Vision. Netherlands Organization for Applied Scientific Research (1972)

53. Schor, C.M., Wood, I.: Disparity range for local stereopsis as a function of luminance spatial frequency. Vis. Res. **23**, 1649–1654 (1983)
54. Ratan, R.: Self-presence, explicated: body, emotion, and identity. In: Luppicini, R. (ed.) Handbook of Research on Technoself: Identity in a Technological Society, p. 322. Information Science Reference, Hershey (2013)

A Non-domain Specific Spatial Ability Test for Gamers Using Drawing and a Mental Rotation Task

Theodor Wyeld[1](✉) and Benedict Williams[2](✉)

[1] Flinders University, Adelaide, Australia
theodor.wyeld@flinders.edu.au
[2] Swinburne University of Technology, Melbourne, Australia
bwilliams@swin.edu.au

Abstract. This chapter describes a study of gamers (defined as persons who play 2D or 3D games for more than 1 hr a week) and non-gamers (defined as those who play 2D or 3D games for less than 1 hr a week/month) and their ability to draw what they see. Participants completed a drawing task, a series of spatial cognition tests, and reported on their gaming habits. Gamers tended to perform better than non-gamers in both the spatial reasoning tests and the drawing task, and statistical analyses showed common processes were involved in both types of tasks. It is likely that similar faculties seem to be invoked by gamers' approach to the tasks when compared to non-gamers.

1 Introduction

Most research in Human Computer Interaction (HCI) and video-games concerns the interface and peripheral devices, player satisfaction, and ease of learning the game-play controls. When discussing the requirement for players' spatial ability (that is, cognitive faculties for representing and manipulating spatial information) to play video-games most research shifts from HCI to cognitive psychology. Cognitive psychology can inform the HCI community by providing an understanding of the role of spatial ability plays for gamers. Most studies suggest that playing video games improves spatial ability, and that gamers have superior spatial ability compared to non-gamers.

A common test used for spatial ability is the Mental Rotation Test (MRT). This is a robust spatial ability test, however, the two core skills involved in game-play are perception and motor skills, and the MRT does not test for motor skills directly. An additional instrument is needed to test for perception and motor skills that respond to the MRT also.

Similar to video-game play, drawing involves both perception and motor skills. It follows then that those who score high on drawing skills should, all else being equal, also score high on gaming skills. Therefore, spatial ability for both drawing and game play can be tested via the MRT. If gamers also score high for drawing and the MRT this would suggest that spatial ability is not restricted to the domain of game-play. Having gamers complete a MRT and a drawing test was used to establish whether gamers' spatial ability is non-domain specific.

© Springer International Publishing Switzerland 2015
T. Wyeld et al. (Eds.): OzCHI 2013, LNCS 8433, pp. 114–132, 2015.
DOI: 10.1007/978-3-319-16940-8_6

1.1 Spatial Skills Critical in Game-Play

Spatial exploration or traversal is not only critical to gain an understanding of the virtual worlds represented in a video game but it is also critical to game-play [43]. For example, in games such as Grand Theft Auto, Fable, and Civilization III the space of the game world is an object of activity. Exploration and navigation of this space is a central aspect of play, with the player often required to cover large distances in order to complete game objectives. Spatial contradictions arise when restrictions are discovered, such as players getting lost (i.e., the game world does not match the player's perception/construction of it), the spatial domain is constrained by the game's map, or the player needs to complete a mission to move to the next level of the map. Other spatial contradictions include the ability for players to pass through otherwise solid objects, such as when using the "no clipping" cheat in Half-Life 2. Players can thus traverse any dimension and become lost outside the normal parameters of the game world.

Most studies investigating the spatiality of video-games come from researchers in the cognitive sciences [7, 8, 12, 16, 21, 22, 50]. HCI studies of video games tend to focus on the interface only. Few studies investigate game-play directly. Most studies of video game interaction conclude that it is quite different to the usual interactivity investigated by traditional HCI [5, 28, 59]. For example, while Nielsen's [40] five usability principles – learnability, efficiency, memorability, error prevention, and satisfaction – can be applied to games evaluation, their usability heuristics are often inverted by game-play. Video games can: be deliberately difficult to learn (as a device to maintain play interest); involve inefficient solutions to problems (making players cover vast distances to complete missions); test the player's memory (players memorise explicit details which are periodically tested); include intentional errors (resulting in penalties for players when they mistime a jump, set off a bomb, fall to their virtual death, and so on). The only one of Nielsen's principles that applies to both games evaluation and traditional HCI is satisfaction. The most common approach has been to investigate game-play interaction using heuristic design guides to evaluation which are based on the designer's professional experience [28, 33, 48]. It is only recently (since the establishment of Microsoft's Playtest Group) that a consistent industry-validated method for video game evaluation has emerged (http://www.microsoft.com/en-us/playtest). Despite this, the spatiality of the game is often overlooked in favour of other factors (playability, type of play, narrative, etc.). Hence, a study is needed of the spatiality of the game – or the players' spatial abilities – if we are to understand the role of space in a game and how this can be supported by HCI.

Gamers' superior spatial ability. Some empirical studies have suggested that gamers may develop enhanced visual attention compared to non-gamers as a result of their video-game playing [8, 21, 22]. Many action video games place greater demands on visual attention than routine daily activities. For example, simultaneous processing of multiple item locations and trajectories is common in video-game play. This also requires rigorous rejection of irrelevant objects – proficient attentional selection processes. Game-play requires objects in view to be matched by their representation stored in memory grounded by a familiar frame of reference – the horizon, or a gravitational grounding. Comparisons between percept and internally stored representations are normalised about the vertical axis according to this gravitational frame of reference [58]. Gamers are often found to have superior spatial skills compared to non-gamers [21] and are thought to develop proficient selection processes because they spend a lot of time engaged in challenging spatial activities.

As gamers play more games and accrue experience with challenging spatial tasks, their abilities in this domain may increase over time. For example, gamers develop a profound understanding of how simulated environments are structured and can be navigated. For simulated 3D environments, this involves an understanding of the underlying principles of perspective: relative size, convergence of parallel lines, occlusion, and so on. Gamers' also develop superior visual attention capacity as evidenced by the way gamers are able to maintain better performance on spatial ability tests as the tests become harder, whereas the non-gamers performance deteriorates rapidly [21]. Hence, gamers' superior visuospatial ability appears to be closely linked to the activity of playing games. Game-play involves visual analysis and evaluation in order to carry out motor execution. Gamers may use the same visual processes in game-play as they do in just observing simulated spatial environments. However, certain visual processes may be crucial only for game-play.

The improvement or enhancement of gamer's cognitive abilities over time may be similar in nature to domain-specific enhancements seen in other types of experts. For example, expert chess players can accurately recall the positions of pieces from a game in play having only seen the pieces for a few seconds [9, 11]. But when the pieces are randomly distributed around the board (i.e., a configuration that would not happen in a real game) the expert chess player's ability to recall the arrangement diminishes to the level of a novice chess player. Novice chess players display the same ability to recall the arrangements in either scenario. This suggests that the expert chess players' ability to recall arrangements is limited to situations they are already familiar with. Their ability is domain specific - the skill of the chess player is restricted to the domain of their expertise – not a superior visual ability more generally. This view suggests that gamers may also be restricted to expertise within their domain. In particular, the 3D gamers' domain being that of a perspective view which when violated may render their expertise similar to the non-gamer. To test this domain specificity, a task is needed whereby their spatial ability manifests in a different form. Drawing what one sees is a spatial task similar to game-play – requiring an acute understanding of perspective cues and how to create their representation motorically – but outside the game-play domain. If gamers' expertise is restricted to game-play alone then they should not show proficiency in other visuo-spatial tasks, such as drawing.

1.2   Gamers, Drawers and the MRT

Video-game play is an activity enjoyed by many. The playing of video-games requires development of certain perceptual and motor skills [21]. Drawing is another activity enjoyed by many in society. Accomplished drawing requires superior perceptual and motor skills. The question arises as to whether the skills required for these tasks overlap? To explore the possibility of general processes underlying both activities, we can test the common contributing factor: visuo-spatial ability and how this correlates with skilled performance – the ability to represent, transform, generate, and recall symbolic, non-linguistic information [26, 31]. To-date, one of the most robust tests for visuo-spatial ability is the Mental Rotation Test (MRT). Although the MRT does not test for motor skills directly it has been reported to involve motoric activation– participants often act out mental rotations by physically indicating turns with their hands or body [44, 56]. Perhaps, then, those who achieve a high score for speed and accuracy in the MRT should also have the necessary perceptual and motor skills to play video games and draw. This study investigates the MRT data for video-game players and drawers to see if there is a correlation. If there is a correlation then combining the MRT and drawing forms a common test for visuo-spatial ability across both domains.

Gamers as Drawers. The drawer attempts to represent a three-dimensional object on a two-dimensional surface. The gamer does not represent three-dimensional objects on a two-dimensional surface directly, rather, the gamer must manipulate their viewpoint in the scene to generate an orientation that advantages their next move. The scene is represented on the monitor screen. In this sense, gamers need to have an acute awareness of the spatiality of the environment they are manipulating to generate the correct representation. Similarly, the drawer may move around the objects and spaces they attempt to represent on paper. In so doing, they also require an acute awareness of the spatiality of the environment they are representing.

While gamers develop their visual attention by practicing game-play, drawers develop their perception by drawing what they see. Perceiving and creating require the drawer to grasp core structural attributes or patterns by organising their visual perception of the spaces before them [1]. This aspect of the drawing process is described by psychological theories of perception that process visual scenes by breaking down the objects into shapes. In terms of Marr's [34] 'primitive sketch', Beiderman's [4] geons, or Pylyshyn's [46] 'feature maps', the drawer organises the three-dimensional spaces, shapes, and objects they need to depict on a two-dimensional surface. By isolating the structurally important features of objects in a scene, the drawer becomes better at attending to the necessary variant information of specific objects [19, 25]. In this way, the drawer develops their skill at analysing and understanding spatial environments more generally. Through game-play, gamers also isolate the structurally important features of objects in the scenes they encounter. Both require the recognition and representation of three-dimensional objects which have been projected onto a two-dimensional surface.[1]

---

[1] The three-dimensional objects projected onto a two-dimensional surface (paper or monitor) are grounded by a horizontal or gravitational frame of reference. It is this horizontal grounding, or sense of the vertical and horizontal structure, that assists recognition of the objects depicted. Both gamers and drawers enhance their visual attention as a result of extended practice. This is the same mechanism by which the MRT is resolved. The armatures of the MRT are recognized and mentally rotated based on their matching shapes and orientations stored in memory. Comparisons between percept and internally stored representations are normalised about the vertical axis more easily than the horizontal according to a gravitational frame of reference [58].

Drawing a real object. Although many studies have investigated the spatial ability of video-game players who interact with three-dimensional scenes on a monitor, few have investigated whether gamers' superior performance in virtual spaces can transfer to real 3-D objects. This is also true of studies of drawers and how they represent three-dimensional objects on a two-dimensional surface. Line drawings of 3D objects are more typical [29, 36, 44]. However, Barbour, and Meyer [3] demonstrated that more accurate drawings could be produced when drawers were shown real objects rather than their line-drawn equivalent. Accuracy can be attributed to the way "…only real objects can supply the perceptual cue defined by binocular disparity that is necessary for the visual system to interpret the depth plane in three-dimensional scenes" [47, p1137]. To address this gap, the present study required participants to draw a real object. This ensured that the results captured their spatial ability as it relates not just to virtual environments but the real world also.

The present study evaluated gamers and non-gamers in a series of tasks examining if there is a relationship between spatial perception and drawing. This involved a series of visuospatial tasks to determine standard spatial abilities and a drawing task which required both visuospatial ability and motor skills. If gamers are superior to non-gamers only in the motoric skills which facilitate drawing, then both gamers and non-gamers should be equally proficient in the visuospatial tasks. If the gamers have a genuinely superior visuospatial ability then this should be manifest in both the drawing task and the visuospatial task performance.

1.3   Defining Gamer/Non-gamer

Researchers in the cognitive neurosciences have investigated methods to improve visuospatial ability using games as training tools [10, 39, 54]. One of the most widely used instruments to measure these effects is the MRT (e.g., Vandenberg and Kuse [56]). Although research indicates that video-games are played mostly by males [10, 53] studies have shown that both men and women can improve their spatial task performance by playing video games [15, 35, 51]. Interestingly, these results are not restricted to training using only games which depict three-dimensional scenes. The two-dimensional game Tetris has been shown to improve visuospatial ability as measured by MRT scores. Okagaki and Frensch [42] found that both female and male college students improved their MRT speed and accuracy after playing only six hours of the game Tetris. Similar results were found by De Lisi and Cammarano [12] when participants played the three-dimensional game Blockout. This suggests that it is the resolving of shapes on the screen and visual attention (the cognitive skill showing the largest difference between gamers and non-gamers) that leads to improvements in MRT score, regardless of whether the game is three- or two-dimensional in nature.

Several confounding factors need to be considered when interpreting these results, these include gender performance differences and experience with video-game play and inherent prior ability. Several reports suggest, in general, male participants tend to have had more computer-mediated spatial experiences than females [2, 12, 24, 41, 55, 57]. This includes video-game experiences [12, 14, 23, 38, 45, 52]. This is despite Green and Bavelier [21, p534] demonstrating across "…video-game playing enhances the number of visual items that can be unerringly apprehended" for both genders. A comprehensive study of gamers is yet to be conducted which determines whether video-game players simply have a greater natural visuo-spatial ability and so were successful at playing video games and thus played more often when compared to non-video-game players whose inherent inabilities and subsequent reduced success resulted in them avoiding playing video-games. This also applies to the inherent superior motor skills and attentional skills of video-game players when compared to non-video game players.

Gender difference in preference and experience can be overcome by training with specific game types, as Terlecki and Newcombe [53] report. The stimuli in MRTs can also be modified – for instance, using "organic" and "everyday" objects - to reduce or eliminate the gender differences regularly observed with the traditional MRT using orthogonal shapes composed of cubes (see Foster and Gilson [18]; Sack, Lindner, and Linden, [49]; Waszak, Drewing, and Mausfeld, [58]).

To address the potential confound of pre-existing capabilities in the participant groups Green and Bavelier [21] conducted a study where a control group of non-video-game players was trained using the game Tetris over the same time span (one hour per day for ten days) as the main group, who were then compared to an untrained group drawn from the same population to detect test-retest improvements. They found that all participants improved their scores, suggesting that inherent attentional and motoric skills are not a major factor in visuo-spatial ability.

The studies by Green and Bevalier [21] and Okagaki and Frensch [42] involved training non-video-game players and video-game players to determining what effect training had on their visuo-spatial ability as demonstrated by their MRT score (among other visuo-spatial tests). Relevant to this study is the finding that both three- and two-dimensional game play resulted in improved visuo-spatial ability. Hence, the definition of gamer and non-gamer must come from the frequency and type of game-play as self-reported by the participant population used in the current study, tabled in the next section.

2 Method

The study reported here is a preliminary analysis of some data from a larger study that attempts to tease apart the perceptual and motor skills used in drawing. It identifies gamers and non-gamers as a differentiable subset of participants for statistical analysis purposes. Part of that statistical analysis is included here.

2.1 Participants

Two-hundred-and-seventy-six participants were recruited over a 2 year period. Participants were first year Education students at Flinders University, Australia, enrolled in a Visual Arts course as part of their Education degree. The Visual Arts course is compulsory for all Education students, but, given the choice, not all students would voluntarily enrol in the Visual Arts course. Participants were evaluated at the beginning of the Visual Arts course before any instruction in drawing was provided. Participation in the study was voluntary. Participants were not paid to participate. Participants took at least 1 hr to complete all tasks.

Of the 276 participants included in the study 250 completed all the requirements (166 women and 84 men). From the 250 who completed all of the spatial ability tests there were:

- 67 "serious" 3D gamers (play at least once a week for more than 1 hour) [25F, 42 M],
- 31 "casual" 3D gamers (play at least once a month for more than 1 hour) [24F, 7 M],
- 18 "serious" 2D gamers (play at least once a week for more than 1 hour) [0F, 18 M],
- 17 "casual" 2D gamers (play at least once a month for more than 1 hour) [13F, 4 M] and
- 117 non-gamers (play less than once a month for less than 1 hour) [104F, 13 M].

These can be more generally divided into 85 Gamers (serious 2D and 3D gamers) [25F, 60 M] and 165 non-Gamers (including casual 2D and 3D gamers) [141F, 24 M].

2.2   Tasks

There were two distinct parts to the study: an online survey and computerised spatial ability tests, and a drawing exercise. The online survey asked participants to provide demographic data, self-report their perceived drawing ability, and frequency of 3D and 2D game playing. This was followed by an interactive Fitts [17] test – used as a measure of participants' ability to use a mouse to find and select objects on a screen [13, 30, 32]. Participants then completed a series of spatial ability tests: a Mental Rotation Test [56]; an 'organic' version of the MRT test to eliminate the reported male preference for orthogonal objects [18, 49, 58], and; a perspective taking test – to determine a participant's spatial orientation ability [27]. Participants had 10 minutes to complete each of the online tasks.

The drawing exercise required participants to draw a familiar object (a teapot) from a predefined distance [6]. Drawings were done on a white A4 sheet of paper taped to a tilted 43° drawing board in a landscape orientation at a predetermined distance of 600 mm from the edge of the table. The teapot was on the same table surface as the drawing board.

Online Survey. The online survey asked participants to provide the following information:

- Gender
- Schooling: public/private
- Cultural background
- Age
- What computer games played: 2D, 3D, none
- How often played: more than once a week, once a week, once a month, once a year
- For how many hours: 1, 2, 3, more
- How do you rate your drawing ability?: good, average, poor
- Can you draw an object realistically? yes, no, maybe

- Where did you learn to draw realistically? primary school, secondary school, books/ self taught, none, other
- What makes a drawing realistic? form/shape, shade/shadow, both

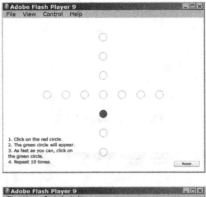

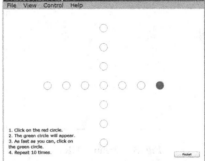

Fig. 1. Fitts test adapted from Dennerlein and Yang [13].

Spatial Ability and Motor Skill Tasks. The Fitts [17] test required participants to point-and-click, using a mouse, on coloured dots as they appear on the screen [13]. This was repeated 10 times. Response time and spatial accuracy for each target was recorded (see Fig. 1.).

The MRT test [56] included 20 pairs of blocks. Half the pairs show identical but rotated block; the other half show non-identical blocks. Participants indicate whether the blocks are the same or different, regardless of rotation. Response time and accuracy were recorded (see Fig. 2).

The MRT organic test [18] follows the same procedure as for the standard MRT. The MRT "organic" was included to overcome the reported male superior performance with the orthogonal MRT (see Fig. 3).

The Perspective Taking test [27] included 20 scenarios. Participants were presented with a scenario, for example: "Imagine you are at the stop sign and facing the house. Point to the traffic light." Participants then moved the dashed line with the arrow head to the pointed position. How long this task took and how accurate participants were was recorded (see Fig. 4).

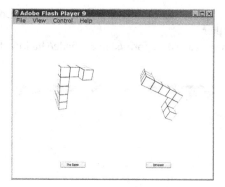

Fig. 2. Standard MRT.

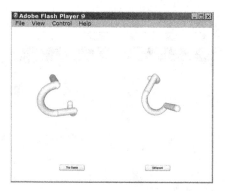

Fig. 3. Organic MRT.

Fig. 4. Perspective-taking test.

Drawing Task. The drawing task required participants to draw a teapot that was positioned on a table in front of them (see Fig. 5). They were told to "draw exactly what you see as accurately as possible – take your time". All drawings were done with pencil and paper. Participants could sketch, erase, and make corrections. Their drawings were

Fig. 5. Seating arrangement for the drawing task.

compared with a photograph taken from the same angle of view as the participant (see Fig. 6). The veridicality of drawings was assessed by a computer program. The operator of the program traces a vector outline of (a photograph of) the object and this was used to determine the percent overlap of this rendering with a similarly traced vector outline of the participant's drawing, after adjusting for scale and orientation (see Fig. 7).

Fig. 6. Photograph of teapot taken from same position as drawer; typical sketch; and, sketch overlayed on photo for accuracy testing.

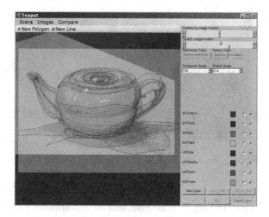

Fig. 7. Screenshot of the software used to determine accuracy of teapot sketch by comparing overlayed sketch on photo and tracing vector lines.

Comparison Across Tasks. To make comparisons across all tasks simpler, task scores were standardised by converting them to z-scores (see Fig. 8):

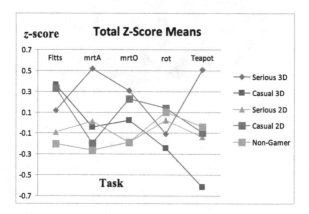

Fig. 8. Comparison of all gamer types and spatial tasks.

3   Results

3.1   Analyses

This section presents a simple analysis of spatial ability differences in gender, gamer/ non-gamer; and, regression analyses of the relationship between spatial ability and motoric performance (drawing test). Comparisons were made between groups of participants and tasks and combinations of tasks. Contrasts between each type of gamer and across gender were made also for each task.

Male/Female Comparisons. Men's and women's performance on each of the tasks is shown in Table 1. Men's and women's scores differed significantly on both the MRT (orthogonal), and the MRT organic. In both cases men's scores were greater, indicating better performance on these tasks, however, the effect was much attenuated for the organic version of the task. Both these findings are consistent with prior research (see Vandenberg and Kuse [56]; Foster and Gilson [18]), although the reasons for both the gender differences and its sensitivity to stimulus type have been debated (e.g., Moe [37]; Terlecki and Newcombe [54]).

Men had a slightly higher drawing accuracy score than women. However, this difference was very small and non-significant.

Women outperformed men slightly (their error was smaller on average) scores on the perspective-taking test, however, this difference was not significant. While perspective taking and mental rotation are often correlated (in this sample there were not r250 = −.05, ns), they are sufficiently dissociable that this lack of significant gender difference is not inconsistent with past findings (see Hegarty and Waller [27]).

Men had slightly higher scores than women on the Fitts task and this difference was significant. This result is likely due to the fact that the majority of gamers (with significant practice and using computer interfaces) are male in this sample.

Table 1. Comparison of Men's and Women's Performance on the Study Measures

Test	Men (n = 84)	Women (n = 181)	t-statistic[a]
Drawing Accuracy[b]	58.11	56.64	t(38.5) = .71,
	(10.72)	(7.31)	p = .48
MRT Cube	72.31	60.37	t(248) = 5.38,
	(17.29)	(16.22)	p < .001
MRT "Organic"	68.48	63.48	t(248) = 2.42,
	(16.83)	(14.75)	p = .02
Fitts task	273.17	257.65	t(137.4) = 2.57,
	(48.11)	(38.21)	p = .01
Perspective taking	38.82	36.19	t(248) = 0.53,
	(29.15)	(39.98)	p = .59

[a]Fractional degrees of freedom indicate a test adjusted for violation of the assumption of homogeneity of variance.
[b]At the time of this publication only 106 of the womens and 31 of the mens' drawing tasks had been analysed.

Non-Gamer/Gamer Comparisons. Participants indicated their gaming habits so we could classify them as "serious," "casual" or non-gamers, and also whether they played predominantly 2D or 3D games, yielding five distinct groups. The number of participants in some classifications (2D gaming) was too small to permit robust statistical comparisons of all gamer types on all measures. Specifically ANOVAs on small and unbalanced group sizes (e.g., relatively few 2D gamers of both seriousness levels, versus the relatively large number of non-gamers) do not yield robust results. Most of the prior research has not distinguished between 2D and 3D gamers' performance on test. On this basis we could argue for collapsing across game type, however, we also performed ANOVAs comparing the serious and casual 2D gamers, and the serious and casual 3D gamers (both these comparisons had balanced groups sizes) and found no significant differences on either mental rotation test, perspective taking, Fitts test, or drawing accuracy. Serious 3D gamers performed slightly better on some tests than some of the other gaming groups, but there were no consistent differences between serious and casual gamers - as defined in this study. Given precedent and lack of significant results we collapse the gaming groups together and compare "gamers" to "non-gamer". In order to justify the separation of non-gamers from gamers by defining gamers as those who play 2D or 3D games more than 1hr per week, and non-gamers who play 2D and 3D games < 1hr/wk, ANOVA for non-gamers and casual 2D and 3D gamers was conducted: $F(1, 163) = 1.63, p > .31$, indicating no difference between those who did not play computer games at all and those who only play less than 1hr per week. Henceforth, casual 2D and 3D gamers were combined with non-gamers in subsequent analyses.

Mean scores on each test are shown for gamers vs non-gamers in Table 2. Gamers showed significantly better scores on both MRTs, however this difference was more pronounced for the cube MRT, or conversely the difference was attenuated for the organic version of the task. This result is congruent with the previously made argument that much of the putative male advantage at cube versions of the MRT arise from experience with orthogonal rotation tasks covertly practiced, for example, in the playing of video games. The same argument applies to gamers regardless of gender – many 3D and 2D games use orthogonal environmental components or platforms which provide a general practice at representing and manipulating such displays.

Gamers showed significantly better performance on the Fitts task, consistent with skills acquired through the playing of computer games. Although it should be noted that the difference was small in absolute terms.

Gamers scored significantly higher on the teapot task, indicating a significant difference between non-gamers and gamers in the ability to draw accurately what they see. This difference was relatively large – about 2/3 of a standard deviation.

Table 2. Comparison of Gamers and non-Gamers Performance on the Study Measures.

Test	Non-Gamer (n = 165)	Gamer (n = 85)	t-statistic[a]
Drawing Accuracy[b]	55.81	60.81	$t(135) = 3.12$,
	(8.09)	(7.37)	$p = .002$
MRT Cube	60.67	71.58	$t(248) = 4.88$,
	(16.42)	(17.34)	$p < .001$
MRT "Organic"	63.51	68.36	$t(248) = 2.35$,
	(14.44)	(17.35)	$p = .02$
Fitts task	261.25	266.01	$t(248) = 0.84$,
	(39.18)	(47.98)	$p = .04$
Perspective taking	38.59	34.13	$t(205.3) = 0.9$,
	(39.12)	(31.33)	$p = .36$

[a]Fractional degrees of freedom indicate a test adjusted for violation of the assumption of homogeneity of variance.

[b]At the time of this publication only 105 of the non gamer's and 32 of the gamer's drawing tasks had been analysed.

The only test in which gamers did not outperform non-gamers was the perspective taking task. In this case gamers still showed a better performance (a lower score indicates a smaller error), but the difference was non-significant.

In sum, gamers showed advantages on the motor test, all but one of the spatial tests, and the drawing (motor and perception).

4 Discussion

The results show that gamers, defined very liberally as those who regularly play more than one hour of video games a week, show somewhat superior performance to non-gamers on the spatial ability tasks (MRTs), and the skilled drawing task, which involves both spatial and motor skill. Performance on the two tasks is highly correlated, and they share similar variances. This congruence could be attributed to common processes underlying both types of task. This further suggests that high level motor skills and spatial ability are required to draw accurately. From these analyses it appears that gamers use at least some of the same spatial skills to both analyse visual stimuli and to represent it in drawing.

To understand difference between non-gamers and gamers performance we need to understand the differences in their visual processing. The authors hypothesise that the gamers' superior performance over non-gamers in both types of task is due not only to their practice at perceiving complex 3D scenes – which seems to underlie the previously reported superiority at MRT – but also the extensive practice at manipulating and navigating simulated spatial environments. Gamers must be able to quickly and correctly perceive and evaluate the objects and shapes depicted on a screen. This is similar to what is required to create an accurate drawing of an object in front of them. Hence, the difference between the cognitive abilities of the non-gamer and gamer may be attributed to partial transfer of the experience and practice the gamer possesses over the non-gamer to these types of tasks. Non-gamers and gamers may use the same procedures to arrive at an accurate drawing. However, the gamer's superior experience provides them with a more accurate and efficient means to put their drawing strategy into action.

In the course of gameplay, gamers are challenged by changing visual stimuli on a screen which they need to make sense of, and in so doing, they need to recognise the objects and shapes depicted. They need to apprehend or construct correspondences between the simulated world and the real world on a regular basis. This is not dissimilar to what one does when they are drawing. In terms of Gombrich's [20] notions of artistic schemata, someone attempting to draw what they see will need to make precise decisions about how to set up the drawing space and how to depict the objects within it. Simple perspective construction, for example, requires knowledge of occlusion, size and distance cues, shading and shadows. The drawer will compare what they see in the world and what they have drawn. In a similar manner, the gamer learns - over time, by interacting with objects in a simulated scene – where objects are in relation to the plane of the screen, their depth cues and so on, from trial and error. In time, the gamer develops a declarative knowledge of the objects and shapes depicted in a simulated environment and a procedural knowledge of how to interact with them. In this sense, the gamers' cognitive knowledge base is different to that of the non-gamers.

5 Implications for HCI

Employing cognitive science methods to investigate the spatial ability of gamers can inform the HCI community of another aspect of game usability. The MRT is a useful instrument in assessing spatial ability in gamers that could also be used in other HCI domains. It is a widely-used and easily administered indicator of spatial cognition. A gamers' spatial ability is critical to their overall enjoyment of the game. The spatiality of a game is often also central to game-play – such as the size of the map, need to traverse space to achieve mission goals, passing through objects, getting lost and so on. Hence, the legibility of the scenes depicted can be a usability issue. In future, evaluations of game satisfaction could also include a measure for spatial comprehension. If it is found to be lacking it may not be the spatial ability of the player at fault, but rather the lucidity of the environments depicted.

Video game interaction is clearly quite different to the usual interactivity investigated by HCI. The usual HCI usability goals are often inverted in game-play – satisfaction being the single enduring goal. Hence, understanding how gamers interact with the space of the game is critical to understanding game-play. As reported, gamers develop superior perceptual and motor skills when compared to non-video-game players. However, by including the drawing task in this study, we were able to demonstrate how gamers' superior perceptual and motor skills might transfer outside the domain of game-play. In turn, this has implications for HCI studies in other spatially-mediated domains, such as aircraft navigation controls, engineering, data visualisation and so on.

As this study demonstrates, gamers seem better able to isolate the structurally important features of objects in a scene, and attend to the necessary variant information (in Gibson's [19] terms) of specific objects than the non-gamer. In so doing, they are also more highly skilled at analysing and understanding spatial environments more generally. That this translated into an ability to also draw a realistic object – the teapot – was surprising. This HCI investigative task, incorporating game-play and drawing, was designed to test perceptual skills combined with motor skills. In so doing, it established a performance predictor for spatial ability. The predictor can be used in HCI and the cognitive sciences.

Computer input devices and their usability parameters are well understood in HCI, however, this is not the case for direct input devices and their effectiveness in virtual environments such as video-games. This study sheds some light on the importance of the cognitive sciences in informing the less well-researched areas in HCI of games and usability. This study shows how a test for spatial ability can transcend the games domain – a significant step towards understanding the HCI constraints of games and potentially spatially-mediated interfaces more generally.

6 Conclusion

The gamers analysed in this study included those who play 2D and 3D games or no games [21]. We found the same reported differences. However, there are some very large differences. For example, comparisons between the standard MRT and teapot scores stand out for the serious 3D gamers. Arguably, 3D games provide the greatest benefit for the standard MRT cube rotation which is a 3D task. This assumes familiarity with the objects depicted in the standard MRT – otherwise a similarly large difference for serious 3D gamers would be expected for MRT organic. The standard MRT and what serious 3D gamers do is similar to drawing what they see – which involves 3D perception, representation and transfer to a page.

As mentioned in the introduction, the drawer perceives and creates by grasping core structural attributes or patterns of the space around them by organising their visual perception in specific ways [1]. They break the objects down into shapes in a scene [4, 34, 46]. The drawer then organises the spaces, shapes, and objects on their canvas. The gamer follows a structurally similar program. By structurally organising the important features of objects and shapes in a scene, both drawer and gamer are better able to attend to the variant information of specific objects [19, 25]. From this study we see that they both become more skilled at analysing and understanding simulated and created spatial environments.

As Gibson [19] points out, the visual system is reliant on understanding the structure of variants and invariants in view. In this sense, gamers have developed superior visual cognition. The gamers' superior visual cognition arises due to their long term attention to spatial variants. They learn this through problem solving interaction with simulated spatial environments. This is distinctly different to superior ability in other domains. The chess expert [9, 11], for example, stores collections of patterns. They do not systematically organise their declarative knowledge like the gamer. Equally, the procedural knowledge (e.g., if this obscures that then I know this is in front of that) that the gamer possesses through practice allows them to analyse novel situations and find solutions – unlike the chess expert who must rely on pre-existing patterns for solutions (according to Chase and Simon [9]). Gamers compile – and through practice internalise/automate – procedures for dealing with novel situations and hence become more proficient at solving them. This appears to be transferable to creating an accurate representation of a familiar object in their view as a drawing.

Unlike chess, the video-games gamers' play never plays out in exactly the same manner every time. A computer game is not as well defined as a chess game. Instead, a flexible procedural knowledge is necessary. Unlike chess, which is mostly a symbolic representation of logically defined procedures, the computer game is less predictable – the logic underlying it is physically constrained by the virtual world in which it is set. In this sense, the gamer needs to develop good visual cognitive procedures to understand the structure of the scenes, shapes, and objects depicted in them.

This study suggests that there is some overlap in cognitive processes required for video-game play and drawing. Further, the processes appear to be common to gamers and non-gamers alike. The gamers' superiority appears to arise simply from their greater proficiency in using the procedures available to them – due to practice and experience at video-game playing.

References

1. Arnheim, R.: Art and Visual Perception: A Psychology of the Creative Eye. University of California Press, Berkeley (2004)
2. Baenninger, M., Newcombe, N.: Environmental input to the development of sex-related differences in spatial and mathematical ability. Learn. Individ. Differ. **7**, 363–379 (1995)
3. Barbour, C.G., Meyer, G.W.: Visual cues and pictorial limitations for computer generated photo-realistic images. Vis. Comput. **9**, 151–165 (1992)
4. Beiderman, I.: Aspects and extensions of a theory of human image processing. In: Pylyshyn, Z.W. (ed.) Computational Processes in Human Vision: Interdisciplinary Perspectives. Ablex, Norwood (1988)
5. Blythe, M.A., Overbeeke, K., Monk, A.F., Wright, P.C.: Funology: From Usability to Enjoyment. Kluwer Academic Publishers, Dordrecht (2003)
6. Buck, J.: The H-T-P techniques: a qualitative and quantitative scoring manual. J. Clin. Psychol. **4**(4), 317–396 (1948)
7. Case, D.A., Ploog, B.O., Fantino, E.: Observing behavior in a computer game. J. Exp. Anal. Behav. **54**(3), 185–199 (1990)
8. Castel, A.D., Pratt, J., Drummond, E.: The effects of action video game experience on the time course of inhibition of return and the efficiency of visual search. Acta Psychol. **119**, 217–230 (2005)
9. Chase, W.G., Simon, H.A.: Perception in chess. Cogn. Psychol. **4**, 55–81 (1973)
10. Cherney, I.D.: Mom, let me play more computer games: they improve my mental rotation skills. Sex Roles **59**(11–12), 776–786 (2008). Springer Science + Business Media
11. De Groot, A.: Thought and Choice in Chess. Mouton, Holland (1978)
12. De Lisi, R., Wolford, J.L.: Improving children's mental rotation accuracy with computer game playing. J. Genet. Psychol. **163**(3), 272–282 (2002)
13. Dennerlein, J.T., Yang, M.C.: Haptic force-feedback devices for the office computer: performance and musculoskeletal loading issues. Hum. Factors **43**(2), 278–286 (2001)
14. Dominick, J.: Videogames, television violence, and aggression in teenagers. Communication **34**, 136–147 (1984)
15. Dorval, M., Pepin, M.: Effect of playing a video game on a measure of spatial visualization. Percept. Mot. Skills **62**, 159–162 (1986)
16. Feng, J., Spence, I., Pratt, J.: Playing an action video game reduces gender differences in spatial cognition. Assoc. Psychol. Sci. **18**(10), 850–855 (2007)
17. Fitts, P.M.: The information capacity of human motor systems in controlling the amplitude of a movement. J. Exp. Psychol. **47**(6), 381–391 (1954)
18. Foster, D.H., Gilson, S.J.: Recognizing novel three-dimensional objects by summing signals from parts and views. Proc. R. Soc. London, Biol. Sci. **269**, 1939–1947 (2002)
19. Gibson, J.J.: The Ecological Approach to Visual Perception. Houghton Mifflin, Boston (1979)
20. Gombrich, E.H.: Art and Illusion: A Study in the Psychology of Pictorial Representation. Princeton University Press, Princeton and Oxford (2000)
21. Green, G.C., Bavelier, D.: Action video game modifies visual selective attention. Nature **423**, 534–537 (2003). Nature Publishing Group
22. Green, G.C., Bavelier, D.: Effect of action video games on the spatial distribution of visuospatial attention. J. Exp. Psychol. Hum. Percept. Perform. **32**(6), 1465–1478 (2006). The American Psychological Association

23. Greenfield, P., Brannon, G., Lohr, D.: Two-dimensional representation of movement through three-dimensional space: The role of videogame expertise. J. Appl. Dev. Psychol. **15**, 87–103 (1994)
24. Grimshaw, G., Sitarenios, G., Finegan, J.: Mental rotation at 7 years: relations with prenatal testosterone levels and spatial play experiences. Brain Cogn. **29**, 85–100 (1995)
25. Hagen, M.A.: The Perception of Pictures: Alberti's Window: The Projective Model of Pictorial Information, vol. 1. Academic Press, New York (1980)
26. Halpern, D.F.: Sex Differences in Cognitive Abilities, 3rd edn. Erlbaum, Hillsdale (2000)
27. Hegarty, M., Waller, D.: A dissociation between mental rotation and perspective-taking spatial abilities. Intelligence **32**(2), 175–191 (2004)
28. Jørgensen, A.H.: Marrying HCI/usability and computer games: a preliminary look. In: Nordchi 2003: Proceedings of the Third Nordic Conference on Human-Computer Interaction, pp. 393–396. ACM Press (2004)
29. Kaushall, P., Parsons, L.M.: Optical information and practice in the discrimination of three-dimensional mirror-reflected objects. Perception **10**, 545–562 (1981)
30. Liao, M.-J., Johnson, W.W.: Characterising the effects of droplines on target acquisition performance on a 3-D perspective display. Hum. Factors **46**(3), 476–496 (2004). Human Factors and Ergonomics Society
31. Linn, M.C., Petersen, A.C.: Emergence and characterization of sex differences in spatial ability: A meta-analysis. Child Dev. **56**, 138–151 (1985)
32. MacKenzie, I.S., Buxton, W.: Extending fitts' law to two-dimensional tasks. In: Proceedings of Computer-Human Interaction CHI 1992 (1992)
33. Malone, T.W.: Heuristics for designing enjoyable user interfaces: lessons from computer games. In: CHI 1982: Proceedings of the 1982 Conference on Human Factors in Computing Systems, pp. 63–68. ACM Press (1982)
34. Marr, D.: Vision: A Computational Investigation into the Human Representation and Processing of Visual Information. MIT Press, Cambridge (2010)
35. McClurg, P.A., Chaillé, C.: Computer games: environments for developing spatial cognition? J. Educ. Comput. Res. **3**, 95–111 (1987)
36. McWilliams, W., Hamilton, C.J., Muncer, S.J.: On mental rotation in three dimensions. Percept. Mot. Skills **85**, 297–298 (1997)
37. Moe, A.: Are males always better than females in mental rotation? Learn. Individ. Differ. **19** (1), 21–27 (2009)
38. Morlock, L., Yamanaka, E.: Measuring women's attitudes, goals, and literacy toward computers and advanced technology. Educ. Technol. **25**, 12–14 (1985)
39. Newcombe, N.S.: Taking science seriously: straight thinking about spatial sex differences. In: Ceci, S.J., Williams, W.M. (eds.) Why Aren't There More Women in Science: Top Researchers Debate the Evidence. American Psychological Association, Washington, DC (2007)
40. Nielsen, J.: Usability Engineering. Morgan Kaufmann Publishers, San Francisco (1994)
41. Nordvik, H., Amponsah, B.: Gender differences in spatial abilities and spatial activity among university students in an egalitarian educational system. Sex Roles **38**, 1009–1023 (1998)
42. Okagaki, L., Frensch, P.A.: Effects of video game playing on measures of spatial performance: gender effects in late adolescence. J. Appl. Dev. Psychol. **15**, 33–58 (1994)
43. Oliver, M., Pelletier, C.: The things we learned on Liberty Island: designing games to help people become competent game players. In: DiGRA 2005: The Digital Games Research Associations 2nd International Conference, Changing Views: Worlds in Play, Vancouver, British Columbia, Canada, 16–20 June 2005 (2005)
44. Parsons, L.M.: Imagined spatial transformations of one's hands and feet. Cogn. Psychol. **19**, 178–241 (1987)

45. Peters, M., Laeng, B., Latham, K., Jackson, M., Zaiyouna, R., Richardson, C.: A redrawn Vandenberg and Kuse mental rotation test: different versions and factors that affect performance. Brain Cogn. **28**, 39–58 (1995)
46. Pylyshyn, Z.: The role of location indexes in spatial perception: a sketch of the FINST spatial-index model. Cognition **32**, 65–97 (1989)
47. Robert, M., Chevrier, E.: Does men's advantage in mental rotation persist when real three-dimensional objects are either felt or seen? Mem. Cogn. **31**(7), 1136–1145 (2003)
48. Rouse, R.: Game Design Theory and Practice. Wordware Publishing, Plano (2001)
49. Sack, A.T., Lindner, M., Linden, D.E.J.: Object- and direction-specific interference between manual and mental rotation. Percept. Psychophys. **69**(8), 1435–1449 (2007)
50. Sims, V.K., Mayer, R.E.: Domain specificity of spatial expertise: the case of video game players. Appl. Cogn. Psychol. **16**, 97–115 (2002)
51. Subrahmanyam, K., Greenfield, P.M.: Effect of video game practice on spatial skills in girls and boys. J. Appl. Dev. Psychol. **15**, 13–32 (1994)
52. Subrahmanyam, K., Greenfield, P.: Effect of videogame practice on spatial skills in girls and boys. In: Greenfield, P., Cocking, R. (eds.) Interacting with Video, pp. 95–114. Ablex, Norwood (1996)
53. Terlecki, M.S., Newcombe, N.S.: How important is the digital divide? The relation of computer and videogame usage to gender differences in mental rotation ability. Sex Roles **53** (5/6), 433–441 (2005)
54. Terlecki, M.S., Newcombe, N.S., Little, M.: Durable and generalized effects of spatial experience on mental rotation: gender differences in growth patterns. Appl. Cognit. Psychol. **22**, 996–1013 (2008). Wiley InterScience, Hoboken (2007)
55. Tracy, D.: Toy-playing behavior, sex-role orientation, spatial ability, and science achievement. J. Res. Sci. Teach. **27**, 637–649 (1990)
56. Vandenberg, S.G., Kuse, A.R.: Mental rotations: a group test of three-dimensional spatial visualization. Percept. Mot. Skills **47**, 599–604 (1978)
57. Voyer, D., Nolan, C., Voyer, S.: The relation between experience and spatial performance in men and women. Sex Roles **43**, 891–915 (2000)
58. Waszak, F., Drewing, K., Mausfeld, R.: Viewer-external frames of reference in the mental transformation of 3-D objects. Percept. Psychophys. **67**(7), 1269–1279 (2005)
59. Zaphiris, P., Ang, C.S.: HCI issues in computer games. Interact. Comput. **19**(2), 135–139 (2007)

Differentiating Cognitive Complexity and Cognitive Load in High and Low Demand Flight Simulation Tasks

Jemma Harris[1(✉)], Mark Wiggins[2], Ben Morrison[1], and Natalie Morrison[1]

[1] Australian College of Applied Psychology, Sydney, Australia
jemma.harris@acap.edu.au
[2] Macquarie University, Sydney, Australia

Abstract. In the contemporary workplace, the design of interfaces has a significant impact on the cognitive demands experienced by operators. Previous approaches to the assessment of these designs have relied on measures of cognitive load to infer the level of cognitive demand imposed. Assessments of cognitive complexity may offer a complimentary measure of the demands of the task as they take into account the inherent nature of the task, rather than idiosyncrasies of the operator. Two studies are reported that examined the information acquisition behavior of pilots in response to a series of simulated flight sequences involving different levels of cognitive complexity. Information acquisition was recorded using an eye tracker. Taken together, the results suggest that assessments of the complexity of a task should be employed as a benchmark in task assessment.

Keywords: Cognitive complexity · Cognitive load · Eye movement · System design · Aviation

1 Introduction

Contemporary approaches to task assessment generally adopt a user-referent perspective in which a range of different users interact with a task and their perceptions and performance are recorded and distributed [1, 2]. The intention is to establish the limits to performance amongst prospective users. In doing so, strategies can be developed that either alter the nature of the task and/or restrict the task to users with particular capabilities or experience. This will pose implication for task design.

The design of tasks around the limitations of users reflects an underlying assumption that the performance of a task imposes a mental or physical load [3, 4]. Depending upon the nature of the task, the load will accumulate to a point where performance begins to deteriorate. This is the point at which the demands of the task exceed the capabilities of the user. At a cognitive level, the demands imposed by a task are presumed to be reflected in perceptions of mental or cognitive load which, in combination with the motivation to maintain a specified level of performance, is associated with the effort invested in the performance of task [5]. Perceptions of cognitive load will be further

© Springer International Publishing Switzerland 2015
T. Wyeld et al. (Eds.): OzCHI 2013, LNCS 8433, pp. 133–150, 2015.
DOI: 10.1007/978-3-319-16940-8_7

moderated by the experience of the operator to a point where, arguably, it is possible to achieve relatively high levels of performance, with relatively little effort and a resultant lower perception of cognitive load [5].

From the system designer's perspective, the difficulty with an assessment based on cognitive load lies in anticipating the minimum requirements necessary to ensure that the least able operator is capable of undertaking the task successfully. This requires some understanding of the complexity of the task, since this represents the minimum level of processing necessary to undertake the task, irrespective of experience or performance shaping factors [6]. While experts might be more rapid and more accurate in their capacity to manage the complexity of the task, the complexity associated with the task remains. Therefore, while perceptions of cognitive load may be variable, depending upon the skills and the motivation of the operator, cognitive complexity is characterized by the nature task itself and therefore, can be established a priori.

This chapter explains how the cognitive complexity of a task can be operationalized as a means of standardizing the cognitive aspects of product development in the future. The chapter will start with a discussion on the notion of cognitive complexity and the relationship between the cognitive and behavioural elements of task performance. Sections 3 and 4 will then present empirical data from two studies where cognitive complexity and information acquisition were examined in an aviation context. The aviation focus represents a domain where the utility of usability testing requires a cognitive focus. Aviation further represents a context where the user experience might be generalized to other domains that evoke demands on information processing resources. A discussion on the outcomes of the empirical investigations will follow. Future research directions will then conclude this chapter.

2 The Notion of Cognitive Complexity

Wood [7] explains complexity as a combination of component, coordinative, and dynamic complexity. Component complexity refers to the number of distinct acts that need to be completed and the number of information cues that are required to be interpreted. Coordinative complexity comprises the form and strength of the association between task inputs and between inputs and task outcomes. Finally, the dynamic aspect of complexity includes the extent to which there are variations in the relationship, whether there are sudden changes in the relationships, and whether these changes are predictable.

The distinction between individual and task attributes in the context of task performance is further conceptualized by Woods [6]. Consistent with Wood [7], Woods [6] conceptualizes cognitive complexity as a combination of four indicators: (1) the time-constraint imposed, (2) the uncertainty associated with performing the task, (3) the inter-relationships between components, and (4) the risk associated with the activity.

In the context of cognitive complexity, the level of dynamism refers to the extent to which the information used to perform the task changes over time. The level of uncertainty is the extent to which an operator can rely on the information, while the interrelationship between components relates to the prospective impact of one task on the performance of other tasks [6, 8, 9]. Finally, the risk associated with a task refers to the combination of the likelihood of performing the task correctly and the implications of performing the task incorrectly [6, 9, 10].

Both Woods [6] and Wood [7] argue that, independent of motivation and task experience, there is an association between the cognitive characteristics of a task and the demands on information processing resources. For example, increases in the rate at which task-related information changes (referred to as dynamism) will require an increased capacity to monitor, extract, and integrate information over shorter periods of time [1, 6]. Similarly, an increase in the number of components with which an element is associated will require an increased capacity to anticipate the consequences of events [6].

Where the indicators of system functioning become inaccurate, a level of uncertainty is generated that requires an assessment of the accuracy of the information, based on either previous experience or on other information available at the time [6, 8]. Finally, increases in the risk associated with an incorrect response to a task are associated with an increase in the demands for the management of affective responses that might reduce the capacity of working memory [1].

2.1 Cognitive Complexity and Task Performance

In establishing the relationship between cognitive complexity and task performance, Maynard and Hakel [4] distinguish objective and subjective conceptualizations of the construct. Their conceptualization of objective complexity broadly reflects the theoretical positions of Wood [7] and Woods [6] insofar as it is defined as the amount of information that is required to be integrated to optimize task performance. By contrast, their conceptualization of subjective complexity is more consistent with notions of cognitive load, since it is defined as the perception of the level of complexity associated with the performance of a task.

Like cognitive load, Maynard and Hakel [4] contend that the level of subjective task complexity reflects the impact of performance shaping factors such as motivation and experience, and moderates the effort ascribed to the performance of the task. Importantly, they found that levels of subjective and objective complexity, although related, contributed differently to performance on a managerial scheduling task. This suggests that it is both the subjective and objective demands of a task that contribute to task success.

Although subjective and objective complexity appear to contribute independently to task performance, the difficulty for system designers lies establishing systematically, the differences in the demands of the task. Where subjective complexity can be established through the administration of questionnaires, objective complexity needs to be established as an outcome of a task analysis. This process can be time-consuming and may, in and of itself, represent a subjective conceptualization of the demands of the task.

An alternative approach to a detailed task analysis involves an assessment of the process of information acquisition as indicative of the objective complexity of a task. Information acquisition and the behavioral strategies with which it is associated are well-established measures of underlying cognitive performance. First-order measures include the data arising from process tracing [11] and visual search [12, 13], while second-order measures include data from retrospective verbal protocols, self-reports [14, 15], and cognitive interviews [16, 17].

As a measure of visual search, fixation duration has been used to draw inferences to cognitive processing, albeit with inconsistent outcomes. For instance, in Tole, Harris, Stephens, and Ephrath [18] average fixation time increased when mental workload was raised, yet Rivercourt, Kurperus, Post and Mulder [19] observed a decline in fixation duration with increasing levels of task complexity. These discrepant outcomes might be explained by the inherent value of average values, and the time frame over which eye movements are analyzed.

The value of assessing the dispersion of fixation durations over time has been raised by Velichkovsky, Dornhoefer, Pannasch, and Unema [20]. In their study, the proportion of fixation durations occurring in five categories was calculated. In the first second following the release of a critical event, fixation durations increased significantly in the >601 ms category. This outcome could be interpreted as providing support for the notion that increases in task complexity are related to longer fixation durations. However, during the fifth second of the critical event, fixation durations were predominately occurring in the <151 ms category [20]. Therefore, it appears that the period in which fixation durations are analyzed is important for making inferences to cognitive processing.

3   Study 1: Task Complexity and Information Acquisition

Wiggins [21] proposes that, under conditions of uncertainty, the pattern of information acquisition during the diagnostic or situation assessment phase of a task reflects the underlying complexity of that task. Tasks that are objectively more complex will require the acquisition of greater amounts of information, with greater frequency, and within less time. By contrast, tasks that are less complex will require the acquisition of lesser amounts of information, with less frequency, and within a greater period of time.

Given that objective complexity is based on the intrinsic demands of the task, it should be possible, having completed the task, for users to rate the various dimensions of complexity, provided that questions relate to the underlying characteristics of the task, and not perceptions of the demands of those tasks. In doing so, designers could be offered a relatively inexpensive solution to the assessment of both subjective and objective complexity in the context of system design.

3.1   Aims and Hypotheses

The present studies were designed to establish whether differences in the objective complexity of tasks are associated with differences in information acquisition during flight simulation; whether these differences are consistent with the effects implicit in Wiggins [21]; whether pilots' ratings of the elements of objective complexity relate to the intrinsic complexity of the tasks; and whether ratings of objective and subjective complexity discriminate between tasks of greater and lesser complexity.

Aviation was employed as a context for the research since the use of flight simulation provides a combination of ecological validity and experimental control. It also constrains the number of features available for problem resolution, and enables the application of eye-tracking technology. Finally, pilots are required to record their experience in operating aircraft, thereby providing a relatively accurate indication of both the quality and the quantity of operational experience.

Study 1 involved the development of a series of flight simulated problem-solving scenarios that that differed objectively in their complexity. On the basis of Wiggins [21], it was hypothesized that, in comparison to objectively complex scenarios, pilots' would access less information, with less frequency, and within more time during the less complex scenarios.

3.2   Method

Subjects. This study was approved by the UWS Human Research Ethics Committee and subject's gave their written informed consent to participate. The subjects comprised 41 general aviation pilots, of whom 38 were male and three were female. They ranged in age from 19 to 62 years, with a mean age of 34 years (SD = 13.19). The subjects had accumulated between 70 and 8000 total flying hours (M = 937.80 h, SD = 1587.10), of which between 10 and 6300 were undertaken as pilot-in-command (M = 683.78 h, SD = 1284.10). The majority of subjects held a commercial or private pilots license, and were neither a flight instructor nor instrument-rated. The subjects represented a convenience sample recruited through a University-based aviation research register and through a number of flight training organizations. They were compensated $40.00 for travel expenses.

Equipment and Materials. The study was conducted using a simulated Cessna 172 aircraft operated on a Precision Flight Controls flight simulator, using the X-Plane 6.21™ program developed by Laminar Research Corporation. Figure 1 displays the layout of the flight simulator. The pilot was seated in a fibreglass cockpit and communicated with the experimenter through a headset. Through the instructor PC, the X-Plane program allowed for the airport, aircraft, weather, time of day, and instrument serviceability to be varied manually. It also contained a dynamic display of the aircraft's altitude, heading and airspeed, and allowed the experimenter to view the location of the aircraft with reference to nearby airfields.

Developed by Seeing Machines, FaceLAB™ Version 3.2 was used to record eye-gaze data. The system uses the input from two small video cameras (located on the flight simulator console) to determine gaze direction and fixation duration. A graphical model containing information on the size, orientation and spatial separation of real-world objects was created to represent the flight simulator environment. Within this graphical model, the flight simulator was divided into seven Areas of Interest (AOI, see Fig. 2). The individual instruments located on the instrument panel (e.g., altimeter) were not mapped onto the model, as a faceLAB AOI can be no smaller than 32×24 cm in actual size and most of the individual flight simulator instruments were 5×10 cm. Nevertheless, the model created for the present study could generate reliable information on the major AOIs that were accessed whilst managing the aircraft. The subjects' gaze interaction with the graphical model (e.g., point of fixation and fixation duration) was logged every 16.66 ms (sampling rate of 60 Hz). The 'xlFAT' add-in software, also developed by Seeing Machines, was used to analyze the gaze data.

Note. 1. outside, 2. airspeed indicator, 3. artificial horizon, 4. altimeter, 5. GPS, 6. magnetos, 7. engine instruments, 8. turn and balance coordinator, 9. directional indicator, 10. vertical speed indicator, 11. RPM, and 12. mixture.

Fig. 1. The geographic location of aircraft instruments in a simulated Cessna 172 aircraft operated on a Precision Flight Controls flight simulator.

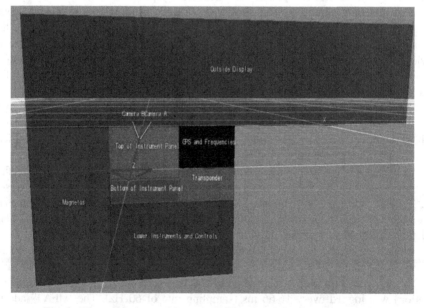

Fig. 2. The seven areas of interest (AOI) created for the present study.

The flight sequences were designed in consultation with five subject-matter experts who were cognizant of Woods' [6] notion of cognitive complexity. The five subject-matter experts (SMEs) were asked to develop three flight sequences that differed systematically from straight and level flight on the basis of four elements of objective complexity, including: (1) the requirement for an accurate diagnosis, (2) the uncertainty associated with task, (3) the number of components required to reach a diagnosis, and (4) the risk associated with a misdiagnosis. The three flight sequences in successively increasing levels of objective complexity included an airspeed indicator failure, a right magneto failure (partial engine failure), and a low oil pressure indicator (the precursor to a total engine failure). The straight and level flight sequence acted as a baseline against which the assessments of performance in response to other flight sequences could be compared.

A questionnaire was administered to the subjects following the completion of each of the four flight events. The questionnaire comprised five questions relating to the objective complexity of the task and was employed as a manipulation check for the purposes of Study 1. Subjects responded to each question on a three-point scale. The summed scores across the five questions ranged from 5 to 15, with a higher score corresponding to a relatively greater level of objective complexity (see Table 1 for the post-flight sequence questionnaire).

Table 1. Post flight sequence questionnaire used in Study 1

Cognitive complexity dimension	Question/s
Dynamism	To what extent did the information that you used to perform the scenario change over time?
Uncertainty	To what extent could you rely on the information that you used to perform the scenario?
Interconnectedness among sub systems	How difficult was the scenario?
Risk	(a) How often did you make mistakes during the scenario? (b) How important was it to avoid errors during the scenario?

Procedure. Having provided written informed consent, the gaze tracker was calibrated and the subjects were instructed to complete a two-minute practice flight to become accustomed to the simulator. Following the practice flight, the experimenter explained the details of the study to the participant. They were instructed that they would be the pilot-in-command of a Cessna 172S aircraft and navigating under visual flight rules. They would complete a take-off, climb and cruise around Tamworth (located in New South Wales, Australia) airport. This airport was selected for its relative novelty to the subjects.

As a context for the scenario, subjects were asked to follow air traffic control instructions in undertaking a visual search for possible bushfires (forest fires). The subjects were naïve as to the exact nature of the study, and particularly the failure of cockpit instruments. However, they were advised that an 'intelligent aviation assistant' would inform them of any abnormalities during the flight by announcing the event that had occurred. The use of an 'intelligent aviation assistant' was explicitly designed to ensure a consistent trigger for the onset of events. The experimenter acted as both air traffic controller and 'intelligent aviation assistant'.

Four flight sequences were completed by the subjects over a period of 20.5 min. Following a period of time during which the participant responded to the failure, the 'intelligent aviation assistant' informed the subjects that the failure had been rectified and the flight had resumed. The straight and level flight sequence was always completed on commencement both to orientate the pilot to the task and to avoid priming. The presentation of the remaining flight sequences was counterbalanced.

3.3   Results

Prior to analysis, the data were screened to ensure that assumptions of normality were satisfactory. Five univariate outliers were identified and the data corrected using a square root transformation. The assumptions of normality were subsequently achieved for univariate and multivariate analyses. Data screening revealed no significant correlation between total hours of flight experience and each of the dependent variables, indicating that total hours of flight experience (as a measure of expertise) did not need to be included as a covariate in subsequent analyses. As a manipulation check, differences in the self-rated objective complexity scores were assessed for the four flight sequences (take off/climb/cruise, airspeed indicator failure, low oil pressure, and right magneto failure) using a one-way repeated measures ANOVA. With alpha set at .05, a statistically significant outcome was observed, $F(3, 120) = 17.58$, $p = .00$, partial $\eta^2 = .305$. Table 2 lists the average self-rated cognitive complexity score (+SD) and outcomes of Bonferroni pairwise comparisons. Consistent with expectations, the right magneto failure and low oil pressure events achieved objective complexity scores that were significantly greater than the scores for the take off/climb/cruise and airspeed indicator failure flight sequences. Finally, unless otherwise stated, for all statistical tests alpha was set at .05.

The number of AOI accessed during the two time-periods (30–0 s prior to the onset of the instrument abnormality, and 0–30 s following the release of the instrument failure) associated with the three instrument failure events were analyzed using a 3 × 2 repeated measures ANOVA. A statistically significant interaction was observed, $F(2, 80) = 8.27$, $p = .003$, partial $\eta^2 = .171$. Table 3 shows the means, standard deviations and significance level for each variable. The pilots acquired a significantly greater number of AOI during the higher objective complexity events in comparison to the lower cognitive complexity events.

To assess the time spent examining the information acquired, fixation durations were divided into five categories: (1) <151 ms, (2) 151–301 ms, (3) 301–450 ms, (4) 451–600, and (5) >601 ms; and the proportion of fixation durations occurring in each category was calculated. As employed by Velichkovsky et al. (2001), this method of assessing fixation duration data is favored over the interpretation of average fixation durations, as measures of central tendency tend to ignore the dispersion of data. Across

Table 2. Descriptive statistics and analysis of variance for self-rated cognitive complexity scores across the four flight sequences (N = 41).

	Take/off, climb, cruise	Airspeed indicator failure	Right magneto failure	Low oil pressure	Significance (2-tailed, post-hoc)
		7.88 (1.34)	9.04 (1.57)		$p = .000^*$
	8.28 (1.18)		9.04 (1.57)		$p = .006^*$
Self-rated cognitive complexity		7.88 (1.34)		9.50 (1.43)	$p = .000^*$
			9.04 (1.57)	9.50 (1.43)	$p = .089$
	8.28 (1.18)			9.50 (1.43)	$p = .000^*$
	8.28 (1.18)	7.88 (1.34)			$p = .069$

*Difference between group means is significant at the .0083 level.

Table 3. Descriptive statistics and analysis of variance for the number of AOI accessed: Prior and post abnormality (N = 41)

	Time period		
Event	Pre failure	Post failure	Significance (2-tailed, post hoc)
Airspeed indicator failure (Lower CC)	5.68 (1.01)	5.51 (1.40)	$p = .489$
Right magneto failure (Higher CC)	5.39 (1.02)	6.17 (.80)	$p = .000^*$
Low oil pressure (Higher CC)	5.14 (1.28)	6.07 (1.82)	$p = .000^*$

*Difference between group means is significant at the .017 level

the subjects, the proportion of the total number of fixation durations occurring in each fixation duration category was calculated for the 30–0 s prior to the onset of each failure event and the 0–30 s following the activation of these failure events.

With alpha set at .017 (.05/3), a two way chi-square revealed differences in the proportion of fixation durations prior to, and post the right magneto failure, χ^2 (4, N = 8145) = 160.67, p = .000 and low oil pressure events, χ^2 (4, N = 8145) = 365.12, p = .000. In the 0–30 s following the onset of the right magneto failure the proportion of fixation durations in the <151 ms category increased significantly from 39.9 % to 60.1 %, while the proportion of fixation durations in the >601 ms category moved from 57.5 % to 42.5 %. Similarly, in the 0–30 s following the release of the low oil pressure failure the proportion of fixation durations in the <151 ms category increased significantly from 37.3 % to 62.7 %, while the proportion of fixation durations in the >601 ms category decreased from 63.45 % to 36.6 %. No significant differences were observed for the 151–301 ms, 301–450 ms, and 451–600 ms categories. Finally, no significant difference was evident for the airspeed indicator failure event, indicating relatively comparable fixation durations prior to, and following this instrument abnormality.

3.4   Outcomes

Study 1 sought to establish whether differences in objective complexity were associated with differences in patterns of information acquisition across a range of simulated failure events. Specifically, it was hypothesized that, in comparison to objectively complex scenarios, pilots' would access less information, with less frequency, and within more time during the less complex scenarios.

As a manipulation check, subjective assessments of objective complexity confirmed the a priori classification of the flight sequences as either more or less objectively complex. Significant differences were observed for the frequency of AOI accessed, and the frequency of longer and shorter fixation durations across the scenarios. Consistent with the hypothesis, a greater frequency of shorter fixations, fewer longer fixations, and a greater number of AOI were accessed during those failure events that embodied greater levels of objective task complexity.

In addition to establishing the relationship between objective task complexity and patterns of information acquisition, the outcomes of Study 1 also suggest that operators, having completed a task, possess the capacity to differentiate flight sequences of greater or lesser objective complexity. However, there was no comparison against measures of subjective complexity. Therefore, it is not clear whether measures of objective complexity account for the perceived cognitive load associated with the tasks or whether they account for the intrinsic characteristics of the task, independent of cognitive load. Study 2 was designed to both replicate and extend the outcomes of Study 1 by incorporating a comparative analysis of measures of subjective and objective complexity.

4   Study 2: Explanatory Power of Task Complexity

From the perspective of cognitive load theory, the resources expended by an operator and therefore, the load imposed, can be established through subjective assessments of task difficulty, while the resources allocated to achieve successful task performance can be established through perceptions of the effort invested [5]. Collectively, subjective assessments of difficulty and effort correspond to elements of Maynard and Hakel's [4] notion of subjective complexity. The aim of Study 2 was to establish which of the operator-rated measures of subjective or objective complexity better discriminates flight simulation tasks representing different levels of objective complexity.

4.1   Method

Subjects. This study was approved by the UWS Human Research Ethics Committee and subject's gave their written informed consent to participate. The subjects comprised 38 general aviation pilots, of whom 34 were male and four were female. They ranged in age from 20 to 63 years, with a mean age of 33 years ($SD = 13.19$). The subjects had accumulated between 50 and 7100 total flying hours ($M = 583.55$ h, $SD = 1138.97$), of which between seven and 6850 were accumulated as pilot-in-command ($M = 436.55$ h, $SD = 1102.41$). The majority of subjects held a commercial or private pilots license, and were neither a flight instructor nor instrument-rated. The subjects represented a convenience sample recruited through a University-based aviation research register and through a number of flight training organizations. They were compensated $20.00 for travel expenses.

Equipment and Materials. The study was conducted using a simulated Cessna 172 aircraft operated on the Precision Flight Controls flight simulator employed in Study 1. The gaze tracker employed in the present study was identical to that employed in Study 1. The study incorporated two flight sequences that corresponded to two levels of objective complexity (higher and lower) based on the outcomes of Study 1. The low objective complexity task comprised a straight and level flight sequence while the high complexity event comprised the failure of the right magneto (partial engine failure). Immediately following the completion of each flight sequence, subjects were asked to complete the post-event questionnaire. The questionnaire comprised six questions relating to elements of objective complexity (see Table 4), and subjects responded to each question on a seven-point scale. Objective complexity rating scores ranged from 6 to 42, with a higher score corresponding to a relatively greater rating of objective complexity. Subjects were also asked to rate from 0 to 100 %, the effort invested during the performance of the two scenarios. Finally, subjects were asked to recall as much of the information that they had acquired during each scenario.

Table 4. Post flight sequence questionnaire used in Study 2

Cognitive complexity dimension	Question/s
Dynamism	To what extent did the information that you used to perform the scenario change over time?
	How many different pieces of information did you need to be aware of to complete the scenario?
Uncertainty	To what extent could you rely on the information that you used to perform the scenario?
Interconnectedness among sub systems	How difficult was the scenario?
Risk	(a) How often did you make mistakes during the scenario?
	(b) How important was it to avoid errors during the scenario?

Procedure. Following the provision of informed consent, the gaze tracker was calibrated and subjects were asked to complete a two-minute practice flight to become accustomed to the simulator. On completion of the practice flight, they were advised that they would assume the role of pilot-in-command of a Cessna 172S aircraft and navigate under Visual Flight Rules (VFR). The scenario involved an approach and landing at Cairns airport on the North-East Coast of Australia.

Subjects were advised that Cairns airport operates under Class C airspace and therefore, they would need to maintain contact with air traffic control using the headset provided. They were also advised that an 'intelligent aviation assistant' would inform them of any abnormalities during the flight. As in Study 1, the use of the 'intelligent aviation assistant' was explicitly designed to ensure a constant trigger for the onset of the failure event. The experimenter acted as both the air traffic controller and the 'intelligent aviation assistant'.

Using the Instructor PC, the experimenter positioned the aircraft 15 miles from Cairns Airport. This ensured that all subjects commenced the flight at the same distance from the runway. The first flight sequence required the management of the aircraft from cruise into a descent profile. The simulation was paused at 10 miles from the aerodrome, and the subjects were asked to complete the relevant post-flight event questions.

In the second flight sequence, the subjects were asked to continue the approach and landing. The right magneto failure was activated when the aircraft was 4.7 miles from Cairns airport and the 'intelligent aviation assistant' informed the subjects of the failure. The second flight sequence was completed once the subjects had landed the aircraft. On landing at Cairns airport, the subjects were asked to complete the final post-flight event questions.

4.2   Results

Prior to commencing the analyses, data were screened to ensure that the assumptions of normality were satisfactory. Across the analyses, seven univariate outliers were observed and corrected for, and assumptions of normality were achieved for univariate and multivariate analyses. Data screening revealed no significant correlation between total hours flight experience and each of the dependent variables, indicating that total hours of flight experience (as a measure of expertise) did not need to be included as a covariate in the analysis. As manipulation checks, differences in self-rated objective complexity, difficulty, and effort associated with the scenarios (cruise/descent, and right magneto failure) were examined using three dependent samples t tests. With alpha set at .017, the right magneto failure scenario was associated with a significantly greater level of objective complexity, $t(37) = 9.10$, $p = .000$, partial $\eta^2 = .691$, perceived difficulty, and perceived effort invested, $t(37) = -8.09$, $p = .000$, partial $\eta^2 = .639$, than the cruise/descent scenario (see Table 5 for descriptive statistics). Finally, unless otherwise stated, for all statistical tests alpha was set at .05.

A dependent samples t test was conducted for the mean number of AOI accessed during the two time-periods (30–0 s prior to the completion of the cruise/descent flight sequence, and 0–30 s following the release of the right magneto failure). No statistically significant difference was evident, $t(37) = -1.17$, $p = .426$, partial $\eta^2 = .036$.

Table 5. Descriptive statistics for self-rated cognitive complexity, perceived difficulty, and perceived effort across the two scenarios (N = 38).

Variable	Scenario	
	Cruise/descent	Right magneto failure
Self-rated cognitive complexity	20.68 (4.90)	29.08 (3.07)
Perceived difficulty	3.16 (1.53)	5.47 (1.18)
Perceived effort	25.27 (12.45)	52.25 (82.24)

A dependent samples t test established whether differences existed between the frequency of different pieces of information that the subjects reported as having acquired during the two flight sequences. The results indicated a statistically significant difference between the number of different pieces of information that were reportedly acquired during the two tasks, $t(37) = -2.87$, $p = .003$, partial $\eta^2 = .182$. Specifically, the mean number of features reported for the right magneto failure event ($\bar{X} = 7.00$, SD = 1.59) was greater than the mean number of features reported for the cruise/descent event ($\bar{X} = 6.21$, SD = 1.73).

The method of calculating fixation durations was consistent with Study 1. A two way chi-square revealed that the proportion of fixation durations differed significantly prior to, and post the right magneto failure, $\chi^2 (4, N = 7600) = 153.30$, $p = .000$. In the 0–30 s following the release of the right magneto failure, the proportion of fixation durations in the <151 ms category increased from 40.2 % to 59.8 %, while the proportion of fixation durations in the >601 ms category decreased from 63.4 % to 42.5 %. No significant differences were observed for the 151–301 ms, 301–450 ms, and 451–600 ms categories.

A direct, discriminate function analysis was used to establish the precision with which the subjective and objective measures of complexity discriminated the cruise/descent flight sequence from the right magneto failure flight sequence. The predictors were: (1) rated objective complexity scores, (2) perceived difficulty scores, (3) perceived effort scores, (4) the number of different pieces of information that were reportedly accessed, (5) the number of AOI accessed, (6) the proportion of fixation durations in the <151 ms category, and (7) the proportion of fixation durations in the >601 ms category. Predictors 5–7 occurred either during in the 30–0 s prior to the completion of the cruise/descent event or during the 0–30 s following the release of the release of the right magneto failure.

Using a conservative alpha of .01, one discriminant function was calculated, with a strong association between groups and predictors, $\chi^2(7) = 68.63$, $p = <.01$, that accounted for 100 % of between-group variability. The loading matrix of correlations between predictors and discriminant functions suggests that the best three predictors for discriminating the first function are, respectively: (1) rated objective complexity, (2) perceived difficulty scores, and (3) perceived effort scores. As evident in Table 6, the standardized canonical coefficients indicate the stability of rated objective complexity, while perceived difficulty may be an unstable outcome.

With the use of a jack-knifed classification procedure for the total sample of 38 pilots, across the two events, 88.2 % of subjects were classified correctly, compared to 50.0 % who would be correctly classified by chance alone. The stability of the classification procedure was confirmed by a cross-validation run, where there was an 86.8 % correct classification rate, indicating a high degree of consistency in the classification scheme.

Table 6. Results of discriminant function analysis variables related to the cruise/descent *(lower CC)* and right magneto failure *(higher CC)* events

Predictor variable	Correlations of predictor variables with discriminant functions *(Standardized canonical coefficients)* 1	Univariate $F(1, 74)$	Wilks' Lamba
Self-rated cognitive complexity	.811 (.735)	80.10*	.480
Perceived difficulty	.668 (-.099)	54.43*	.576
Perceived effort	.595 (.361)	43.08*	.632
< 151 ms fixation duration category	.582 (.375)	41.33*	.642
> 601 ms fixation duration category	-.380 (-.002)	17.62*	.808
Number of AOI accessed	.188 (-.193)	1.09	.985
Self reports of pieces of information accessed	.095 (.281)	4.29	.945
Canonical R	.789		
Eigenvalue	1.647		

Note. * $p < .001$

4.3 Outcomes

The aims of Study 2 were to replicate the outcomes of Study 1 and to examine the relative utility of measures of subjective and objective complexity in discriminating flight sequences that differed in complexity. Apart from the results pertaining to the frequency of fixations across the various AOI, the results confirmed the outcomes of Study 1 with the more complex scenario eliciting a relatively greater proportion of fixation durations in the <151 ms category, and a relatively lesser proportion of fixation durations in the >601 ms category. Differences were also observed in the ratings of objective complexity.

The outcomes of Study 2 also indicated that amongst the variables, ratings of objective complexity best discriminated the two flight sequences and provided a measure of performance distinct from ratings of difficulty and effort. This suggests that objective and subjective measures of complexity contribute differently to the performance of a task and that both measures can be self-rated at a level that discriminated more complex from less complex tasks.

5 General Discussion

The inherent complexity associated with a product or task is a feature that potentially has implications for human performance. Overly complex systems are likely to be error-prone and/or discarded in favor of systems that are relatively less complex. Maynard and Hakel [4], amongst others (e.g., [7]), conceptualize the complexity of a task as a combination of both users' subjective perception of the complexity of the task and the complexity of the task inherent in its execution.

The subjective perception of complexity corresponds to notions of cognitive load and the difficulty that is expected to be encountered in performing a task successfully. The nature of the situation, and the experience and inherent capability of the user determine both perceptions and difficulty, and the subsequent effort that is likely to be invested in the task. Therefore, the difficulty in relying solely on subjective perceptions of complexity lies in the individual differences between users.

The aim of this study was first to establish whether it is possible to differentiate tasks on the basis of objective complexity. Objective complexity relates to the inherent features of the task and the various cognitive and perceptual activities that need to take place to ensure successful performance. In establishing differences in objective complexity, it becomes possible to assess the utility of alternative designs on the basis of the underlying information processing demands, irrespective of factors such as experience or motivation. A series of four simulated flight tasks were developed by subject-matter experts that differed on the four aspects of complexity proposed by Woods [6]. Controlling for task experience, qualified pilots 'flew' the simulated flights and their process of information acquisition was compared during a diagnostic event.

The results indicated that the process of information acquisition differed on the basis of the objective complexity of the task, with more complex tasks associated with a greater frequency of Areas of Interest (AOI) in the 30 s following the onset of the event. More complex tasks were also associated with a lower proportion of fixations at less than 151 ms. The latter effect occurred in both Studies 1 and 2 and, together with a reduction in the proportion of fixations in the >601 ms category of fixations, highlights the behavioral changes that occur with changes in the complexity of a task. These changes corresponded to user-rated assessments of the objective complexity of the task.

A secondary aim of this study was to determine whether subjective and objective complexity contribute differently to the performance of a task and whether self-rated assessments of objective complexity discriminated tasks of greater and lesser complexity. Consistent with expectations, the results indicated that, amongst the outcome variables, self-rated assessments of objective complexity was the variable that best discriminated the tasks, while self-rated assessments of subjective complexity discriminated the tasks to a slightly lesser extent. This suggests that subjective and objective measures of task complexity comprise distinct, but related constructs that contribute to an overall assessment of complexity.

The joint contribution of subjective and objective complexity corresponds to the integrated approach to the assessment of complexity proposed by Maynard and Hakel [4]. However, the outcomes of the present study also suggest that assessments of objective complexity can be self-rated, thereby reducing the potential costs in establishing the objective complexity of the task through the use of subject matter experts or task analyses.

5.1   Limitations and Implications

By establishing the utility of measures of subjective and objective complexity, the flight scenarios developed in the present study inevitably represented extremes as a means of establishing differences between more and less complex tasks. However, there remains a need to determine whether the self-rated measure of objective complexity retains a degree of sensitivity for graduated levels of complexity. This would enable comparisons whereby relatively small changes in the design of a system can be assessed in terms of their impact on the level of task complexity.

In addition to assessments of graduations in complexity, there is a need to establish whether systematic changes in subjective complexity impact either the process of information acquisition or the self-rated assessment of objective complexity. Since objective complexity relates to the nature of the task, variations in the perceived difficulty or effort associated with the performance of a task should occur independently of the intrinsic characteristics of the task.

From the perspective of system design, the outcomes of the present studies offer an opportunity to ensure that the subjective perceptions of users represent the intrinsic complexity of the task. For example, in assessing a new product, a designer may seek the responses of users ranging in experience from the expert to the novice and may find a breadth of subjective perceptions that are not necessarily based on the inherent complexity of the task specified. Experts may rate the tasks as relatively simple while novices may rate the same tasks as particularly difficult. By comparing these data with the data pertaining to objective complexity, it becomes possible to establish a standard against which the complexity of a task can be compared.

5.2   Conclusions and Future Outlook

This research sought to differentiate subjective and objective complexity as distinctive features associated with the overall complexity of a task. Using simulated in-flight events that required a diagnostic response, differences between more complex and less complex tasks were evident which suggested that information acquisition behavior changes as a function of the objective complexity of a task. However, the results also suggested that self-ratings of objective complexity can be employed to establish the intrinsic complexity associated with the task. The outcomes have implications for system design and development in the future. The authors will continue to work with projects examining the relationship between self and objective task complexity and the behavioral elements of task performance.

The current research would be further strengthened by assessing 'primary-task performance' at the level of the individual operator. In particular, it is likely that, where the primary-task has been performed at a less than satisfactory level, deficiencies in operator behavior might be traced to problems with the perception of the cognitive complexity of the task and the pattern of feature acquisition. Therefore, assessing the alignment, or misalignment, between primary-task performance, the perception of cognitive complexity, and the pattern of feature acquisition, would provide further validation of the outcomes arising from the current project.

References

1. Bedney, C., Karwoski, W.: A Systemic-Structural Theory of Activity: Applications to Human Performance and Work Design. Taylor & Francis, Boca Raton (2007)
2. Woods, D.D., Hollnagel, E.: Mapping cognitive demands in complex problem-solving worlds. Int. J. Man-Mach. Stud. **26**, 257–275 (1987)
3. Bach, C., Scapin, D.L.: Comparing inspections and user testing for the evaluation of virtual environments. Int. J. Hum.-Comput. Interact. **26**(8), 786–824 (2010)
4. Maynard, D.C., Hakel, M.D.: Effects of objective and subjective task complexity on performance. Hum. Perform. **10**, 303–330 (1997)
5. Sweller, J., van Merriënboer, J.J.G., Paas, F.: Cognitive architecture and instructional design. Educ. Psychol. Rev. **10**, 251–296 (1998)
6. Woods, D.: Coping with complexity: the psychology of human behaviour in complex systems. In: Goodstein, J., Andersen, H., Olsen, B. (eds.) Tasks, Errors and Mental Models, pp. 128–148. Taylor & Francis, London (1988)
7. Wood, R.E.: Task complexity: definition of the construct. Organ. Behav. Hum. Decis. Process. **37**, 60–82 (1986)
8. Campbell, D.J.: Task complexity: a review and analysis. Acad. Manag. Rev. **13**(1), 40–52 (1988)
9. Reason, J.: Human Error. Cambridge University Press, Cambridge (1990)
10. Williams, M.D., Hollan, J.D., Stevens, A.L.: Human reasoning about a simple physical system: a first pass. In: Gentner, D., Stevens, A.S. (eds.) Mental Models, pp. 131–153. Lawrence Erlbaum, Hillsdale (1983)
11. Wiggins, M.W., O'Hare, D.: Expertise in aeronautical weather-related decision-making: a cross-sectional analysis of general aviation pilots. J. Exp. Psychol. Appl. **1**, 305–320 (1995)
12. Sarter, N.B., Mumaw, R.J., Wickens, C.D.: Pilots' monitoring strategies and performance on automated flight decks: an empirical study combining behavioral and eye tracking data. Hum. Factors **49**, 347–357 (2007)
13. Schriver, A.T., Morrow, D.G., Wickens, C.D., Talleur, D.A.: Expertise differences in attentional strategies related to pilot decision making. Hum. Factors **50**(6), 864–878 (2008)
14. Patrick, J., James, N.: Process tracing of complex cognitive work tasks. J. Occup. Organ. Psychol. **77**, 259–280 (2004)
15. Renshaw, P.F., Wiggins, M.W.: A self-report critical incident assessment tool for army night vision goggle helicopter operations. Hum. Factors **49**, 200–213 (2007)

16. Salmon, P.M., Stanton, N.A., Walker, G.H., Barber, C., Jenkins, D.P., McMaster, R., et al.: What's really going on? Review of situation awareness models for individuals and teams. Theor. Issues Ergon. **9**(4), 297–323 (2008)
17. Wiggins, M.W.: The development of computer-assisted learning systems for general aviation. In: O'Hare, D. (ed.) Human Performance in General Aviation, pp. 153–172. Ashgate, Aldershot (1999)
18. Tole, J.R., Harris, R.L., Stephens, A.T., Ephrath, A.R.: Visual scanning behavior and mental workload in aircraft pilots. Aviat. Space Environ. Med. **53**, 54–61 (1982)
19. Rivercourt, M., Kuperus, M.N., Post, W.J., Mulder, L.J.M.: Cardiovascular and eye activity measures as indices for momentary changes in mental effort during simulated flight. Ergonomics **51**(9), 1295–1319 (2008)
20. Velichkovsky, B.M., Dornhoefer, S.M., Pannasch, S., Unema, P.J.A.: Visual fixations and level of attentional processing. In: Proceedings of the 2000 Eye Tracking Research and Applications Symposium, pp. 79–85. ACM Press, New York (2000)
21. Wiggins, M.W.: Cue-based processing and human performance. In: Karwowski, W. (ed.) Encyclopedia of Ergonomics and Human Factors, pp. 641–645. Taylor and Francis, London (2006)

Spatial Learning

MolyPoly: A 3D Immersive Gesture Controlled Approach to Visuo-Spatial Learning of Organic Chemistry

Winyu Chinthammit[1](✉), SooJeong Yoo[1], Callum Parker[1],
Susan Turland[2], Scott Pedersen[3], and Wai-Tat Fu[4]

[1] Human Interface Technology Laboratory Australia (HIT Lab AU),
School of Engineering and ICT, University of Tasmania, Hobart, Australia
{Winyu.Chinthammit,yoosj,callump}@utas.edu.au
[2] School of Chemistry, University of Tasmania, Hobart, Australia
Susan.Turland@utas.edu.au
[3] Active Work Laboratory, Faculty of Education, University of Tasmania,
Hobart, Australia
Scott.Pedersen@utas.edu.au
[4] Department of Computer Science, University of Illinois at
Urbana-Champaign, Champaign, USA
wfu@illinois.edu

Abstract. Currently, first-year chemistry students learn about three-dimensional molecular structures using a combination of lectures, tutorials, and practical hands-on experience with molecular chemistry kits. We have developed a basic 3D molecule construction simulation, called *MolyPoly*. The system was designed to augment the teaching of organic chemistry by helping students grasp the concepts of chemistry through visualisation in an immersive environment, 3D natural interaction, and audio lesson feedback. This paper presents the results of a pilot study conducted with a first-year chemistry class at the University of Tasmania. Participating students were split into two groups: *MolyPoly* group (no lecturer in the sessions) and traditional classroom group during the four in-semester classroom sessions over a period of two weeks. We present our comparative analyses over the knowledge-based pretest and posttest of the two groups, by discussing the overall improvement as well as investigating the improvement over the test questions with different knowledge difficulty levels and different required spatial knowledge.

Keywords: Organic chemistry · Gesture controlled · Immersive environment · Visuo-spatial reasoning and learning

1 Introduction

Visual representations and physical models are commonplace in science teaching [1, 2]. One of their applications is to enhance the spatial awareness of concepts [1, 3]. In organic chemistry it is important for students to be able to understand the 3D relationship of the molecules to comprehend concepts such as isomerisation, intermolecular

© Springer International Publishing Switzerland 2015
T. Wyeld et al. (Eds.): OzCHI 2013, LNCS 8433, pp. 153–170, 2015.
DOI: 10.1007/978-3-319-16940-8_8

forces and reaction mechanisms [4, 5]. It has been found that when students understand the 3D spatial arrangement of atoms and molecules, how this relates to the 3D and 2D representations printed in textbooks and how to transpose them into a diagram in order to communicate the particular molecule to another chemist, students do better overall in exam or test results, even when the question does not relate to the shape of the chemical [2]. These results suggest that spatial ability is important for learning scientific topics that involve spatial reasoning.

Indeed, spatial ability, defined by Lohman [6] as "the ability to generate, retain, retrieve, and transform well-structured visual images", has often been considered as an individual difference critical for learning advanced scientific and engineering concepts. For example, a recent longitudinal study [7] showed that spatial ability assessed during adolescence was a salient cognitive attribute among those who subsequently go on to achieve advanced educational credentials and occupations in domains of science, technology, engineering, and mathematics (STEM). Another study [8] showed a web-based virtual environment improved the overall spatial ability focusing on mental rotation and spatial visualization, which provided evidence that spatial ability can be improved through learning environments. Psychological studies have also shown that the ability to perform spatial reasoning and problem solving is critical for predicting performance in a wide range of domains [9–11].

Given its importance, researchers have been evaluating how technologies can be involved in instructional methods to improve spatial ability [12–16]. In particular, interactive 3D environments appear promising to provide the needed tools for spatial ability training. Interactive technologies not only allow students to visualize 2D or abstract elements and relations in 3D, they also allow them to directly manipulate the simulated objects using natural gestures and body movements. These characteristics afford more natural visuospatial reasoning as abstract relations or inferences do not need to be represented and transformed internally as mental imagery [17, 18]. Rather, these relations can be offloaded to perception and action as the learner directly manipulates the 3D simulated objects [19]. Given that external representations of simulated objects are constrained by a medium and unconstrained by working memory, inconsistencies, ambiguities, and incompleteness may be reduced in external representations.

Visuospatial reasoning does not only entail visuospatial transformations on visuospatial information. Visuospatial reasoning also includes making inferences from visuospatial information, whether that information is in the mind or in the world [20]. As cognitive tools, immersive 3D systems facilitate reasoning, both by externalizing, thus offloading memory and processing, and by mapping abstract reasoning onto spatial comparisons and transformations. The systems can organize and schematize spatial and abstract information to highlight and focus the essential information.

In addition to visuospatial reasoning, the interactive nature of these systems allows learners to directly manipulate spatial relations to augment the reasoning process. Interaction of mind and body in reasoning has been revealed when people "interact with virtual objects" in gestures. For example, when people describe space but are asked to sit on their hands to prevent gesturing, their speech falters [21], suggesting that the acts of gesturing promote spatial reasoning. The importance of gestures seems to go beyond manipulation of objects in the external environment, as even blind children gesture as they describe spatial layouts [22]. These results suggest that gesture controlled systems have the potential to facilitate visuo-spatial reasoning and learning.

To summarize, research shows that immersive interactive systems have great potential to promote visuospatial reasoning by externalizing complex spatial relations and providing interactive tools for direct manipulation of spatial objects. In this paper, we describe the design of such a system called MolyPoly that helps students explore and learn complex spatial relations in organic chemistry.

2 Motivation

To help students comprehend the spatial environment of the molecule, many different models can be used. Originally these were provided in a textbook, which contained 2D diagrams and pictures of 3D molecules in different formats – ball and stick, ball and cylinders and space filling. Limitations of physical models include the amount of sets available, the number of atoms (balls) and bonds (sticks) as well as the types of molecules that can be made [3].

As computers and ease of software development became widespread, so did the availability of computer-generated models of organic molecules. Developments in computer generated molecules range from merely drawing a 2D diagram through to having a rotating 3D model of a molecule. Some of these applications have overcome the limitations of the physical models – not only those listed previously but also the different bond lengths and atoms sizes, which occur within different bonding systems [3, 5]. One of the other difficulties for students is the ability to mentally picture a 3D molecule from a flat (2D) representation.

Although there has been much investment and research into computer simulations to teach science, there is still more work needed before they become commonplace within classrooms, such as the issue of teacher preparedness to utilize a simulation to teach science concepts [23]. Therefore the teaching tools that are commonly used in a university chemistry unit consist of textbook images, 2D diagrams and plastic models (e.g. balls and sticks) for spatial awareness of the molecules' shape. While these low-cost models are economic teaching aids, they often require explicit instruction and guidance from teachers, as the plastic models do not have enough constraints to guide students to create organic structures based on their 2D diagrams. An interactive system, although more expensive, can have built-in constraints and feedback that allow students to explore the mapping between 2D diagrams and 3D structures without much guidance from a teacher. While these systems are not intended to completely replace teachers, they are promising tools that complement regular instructions by allowing students to explore and learn on their own. This paper describes the evaluation of a chemistry teaching system named MolyPoly, designed to facilitate visuo-spatial reasoning and learning for spatial knowledge content. To preview our results, we found that students who used MolyPoly to learn by themselves showed significant learning effects comparable to students who learned under teacher guidance in a traditional classroom using plastic models. However, before we describe details of MolyPoly, we will first review the use of 3D models to teach chemistry.

3 Related Work

3.1 Current State of Molecular Chemistry Teaching

Chemistry teaching is largely about students visualizing and coming to terms with how molecules are arranged and structured as there can be very complicated molecular structures. Visual representations and physical models are commonplace in molecular science teaching [2, 24]. One of their applications is to enhance the spatial awareness of concepts [1].

Being able to visualize and understand the structures in 3D is difficult for the students to achieve, so to help students comprehend the spatial environment of the molecule many different types of representative models can be used. As mentioned earlier, originally these were provided through a textbook with 2D diagrams and pictures of 3D molecules. Today, physical models are used to assist the student with the visualisation of the 3D structures. These models come in a variety of different sizes and are used to let students touch and clearly visualize the structures.

3.2 Issues with Molecular Chemistry Teaching Practices

The aforementioned physical models or molecular model kits (Fig. 1) used by chemistry students to aid visualisation of molecules were previously identified to have limitations [3]. Other issues with the physical models are that they can become difficult to assemble when creating larger structures and that it can be difficult for students to manually check if the structure created was correctly built [25] as opposed to being automatically checked by a computer program.

Fig. 1. Molecular **Model kit**

A potential issue with teaching methods may also exist according to Erenay's case study [26], which found that the methods of teaching could make a difference as some approaches can hinder students' exploration into the content. In particular, laboratory work was one such method that can limit exploration into certain topics. In Macgrath's study [27], during a study into teaching students special relativity, demonstrated that due to the nature of special relativity, where studying the effects is quite restricted as the study of special relativity only becomes significant at near light speeds, making teaching the subject difficult. Therefore, a computer simulation was created that allowed the students to virtually experience recreations without having to abide by "real-world" constraints. Furthermore, computer simulations were suggested by Noeth [28] as a more viable alternative to physical models as it has been shown that using computer technology in schools has helped improve retention, traditional instruction and student learning efficiency, especially for students that are underachieving.

3.3   Immersive Environment Teaching Systems

Development of computer generated molecules stem from just drawing a 2D diagram to having a rotating 3D molecule both of which are viewed on 2D screens. Some of the earlier applications already overcame the limitations of the physical models such as the different bond lengths and atom sizes, which occur within different bonding systems [5].

Apart from molecular chemistry, there are many successful examples where teaching systems have indeed helped improve the learning environment as well as other classroom aspects for students. A particular example is Minimally Invasive Surgical Trainer (MIST), an immersive virtual reality (VR) system used to teach surgeons laparoscopic tasks. It was found that the system helped improve the performance of training surgeons as they were likely to make fewer errors than the surgeons who trained without the simulator [29]. This could be an indicator of the precision consistency of simulators, as these systems or applications can be created to consistently (and repeatedly) deliver the same lesson and experience. Teaching systems can also be based upon existing simulators or video games and repurposed into adaptions that can suit specific purposes. For example, repurposed video games have been demonstrated to maximize learning efficiency and classroom activity [30, 31].

Furthermore, it has been found that virtual environments experienced by the user through VR can be personalized to the individual learner's needs and depending on the environment the learner is free to make a variety of choices. These environments can also help the learner focus on the task at hand instead of being distracted by what is physically around them [32]. In another example, VR teaching systems are being used to train people in other areas, such as occupational health and safety training for carpentry students who were provided with a virtual environment to experience certain situations that happen on a construction site. The study found a high level of user engagement within the educational 3D computer game environment, which was a key factor for the positive learning results. Moreover, carpentry students preferred to learn using the game, with increased content understanding and retention, which can lead to skill sets being considered more relevant [33]. A high level of user engagement is a result of an active learning and constructivist approach when the participants are actively engaged with the content and construct their own knowledge.

Another important benefit demonstrated in some immersive VR teaching systems is the ability to teach students without a teacher needing to be present. This was demonstrated by a recent study into a VR dance training system which used motion capture technology to capture motions from the user and then gave real-time on-screen feedback to the users. The results from their experiment found that participants that used the system performed better than their control group (learning by watching videos) and that the participants found the method of learning more interesting and highly engaging than simply learning from watching videos [34].

4 Design of *MolyPoly*

MolyPoly is a basic 3D molecule construction simulation, specifically designed to help first-year university chemistry students learn organic chemistry. The organic chemistry requires students to understand the 3D spatial arrangement of atoms and molecules and how this relates to 3D/2D representations in conventional chemistry textbooks. As discussed earlier, an immersive 3D environment is a potential tool for spatial ability training and gesture-controlled systems have potential to facilitate visuo-spatial reasoning and learning. As a result, the first key design of MolyPoly was to combine these two considerations and create an immersive 3D environment that a user (student) can interact with the molecules using their gestures (enabling by a Kinect device).

Fig. 2. MolyPoly in use (notice Kinect at the bottom of the middle screen).

Figure 2 shows the system running an immersive 3D environment on three wall-size large displays (the HIT Lab AU's "VisionSpace", wall-size rear projection screens). Furthermore, MolyPoly utilized VisionSpace's 3D stereoscopic capability (passive 3D with polarized glasses) to deliver the immersive 3D environment where users could see the depth inside the molecule structures. The second key design of MolyPoly was to incorporate a built-in lesson feedback that a user can access during lessons. This helps make MolyPoly a self-driven learning tool.

MolyPoly contains five different lessons in which participants used to create five molecular organic structures as illustrated in Fig. 3. Each lesson contains one molecule lesson (Methane, Propane, 1-Propanal, 2-Propanal, Methoxymethane).

A user can engage the built-in lesson feedback during the lessons via a Check-progress virtual bottom inside the 3D environment, see Fig. 5. Once engaged, Moly-Poly will determine the structural correctness of the molecule structure that was

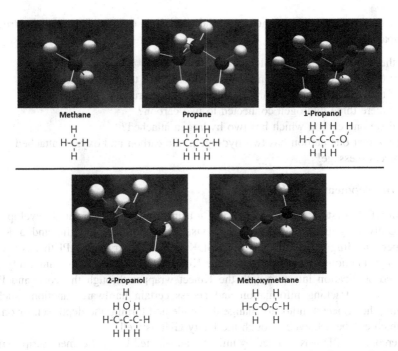

Methane   Propane   1-Propanol

```
     H            H H H            H H H  H
     |            | | |            | | |  |
  H-C-H        H-C-C-C-H        H-C-C-C-O'
     |            | | |            | | |
     H            H H H            H H H
```

2-Propanol   Methoxymethane

```
      H                        H   H
    H O H                      |   |
    | | |                  H-C-O-C-H
  H-C-C-C-H                    |   |
    | | |                      H   H
    H H H
```

Fig. 3. 3D molecules with the corresponding 2D diagrams (Color figure online).

constructed and subsequentially provide audio and visual feedback as to why the structure is correct or incorrect (e.g., there are not enough carbon molecules). The feedback is determined by an algorithm that runs through a set of conditions that need to be met for that particular molecule (lesson). If all of the conditions are met then the application will provide an on-screen success message, as illustrated in Fig. 4, with a 2D diagram of the structure and make an audible announcement of the success. However, if a mistake was made the system will also inform the user through an audible announcement. In addition, the application will provide audible feedback to let the user know the reason. Due to the simplicity of the lessons in our study, we were able to simplify the condition-checking algorithm to be merely a series of one-condition-at-a-time checking and once the fail condition was detected, the algorithm stops and provides the message mostly relevant to the failed condition.

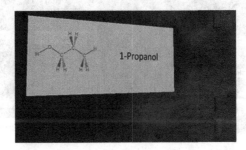

Fig. 4. Visual success message displayed on the left screen of the VisionSpace (Color figure online)

Below is an example of the sequential conditions used in the molecule structure '1-Propanol':

1. Is there one oxygen (red colour ball)?
2. Are there one carbon and one hydrogen connected to one oxygen?
3. Are there a total of three carbons existing in the structure?
4. Are there three hydrogen connected to two carbon?
5. Is there one carbon which has two hydrogen attached?
6. Check that one carbon has two hydrogen, one carbon and oxygen attached.
7. Success message!

4.1 Development of *MolyPoly*

The MolyPoly system was developed in C# using Unity 3D 4 for rapid development. Two Unity plugins were PlayMaker, a visual programming plugin, and a Kinect Wrapper, enabling access to the Kinect SDK. The wrapper is an API that is used to access the functions that are present in the Kinect SDK version 1.7 within Unity. The compiled application interacts with the Kinect wrapper through the code and Play-Maker to get tracking information and access certain hardware functions such as adjusting the Kinect's motor to change the angle and turning the depth sensor on and off, which can be achieved through the Unity GUI as well.

Stereoscopic 3D was enabled by utilizing an adapted Unity 4 camera script written in C# that enabled the application to be viewed in 3D using 3D glasses which allows the users to experience a "pop-out" effect where the molecules feel as if they are coming out of the screen and into physical space. This script was created by the HIT Lab Australia specifically for student projects to utilize the VisionSpace screens' stereoscopic functionality (http://www.hitlab.utas.edu.au/wiki/Stereoscopic_3D).

4.2 Interactions

Gestures were detected through the use of the Kinect skeletal tracking. The skeletal tracking can detect the positions of two hands and in turn, they are represented in the immersive 3D environment as cartoon-style 3D hand avatars, as illustrated in Fig. 5.

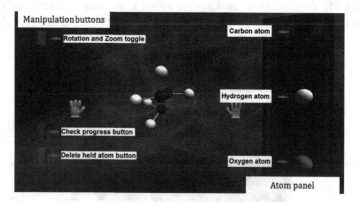

Fig. 5. MolyPoly's immersive workspace with description annotation (Color figure online)

The workspace of MolyPoly is displayed on the VisionSpace projection screens with the molecule structure that is under construction displayed at the center of the workspace and features, which consist of two main control panels: (1) manipulation panel (on the left) and (2) atom selection panel (on the right), as detailed in the following:

Manipulation panel (left side of the application):

- The (top) blue button enables the rotation and zoom.
- The (Second from top) green button gives audible feedback on the correctness of the molecule structure that the user is constructing.
- The (bottom) red button deletes an atom that is currently held in either virtual hand.

Atom panel (right side of the application):

- This area allows the user to pick up carbon, hydrogen or oxygen atoms for connecting with the main structure in the center of the workspace. (Note that only one atom can be held at a time).

All buttons can be activated by simply hovering the virtual hand over them for two seconds (to avoid random hand movements accidentally triggering the buttons). However, the two panels make available two different sets of interaction. The manipulation panel enables users to rotate and scale (zoom) the 3D molecules, while the atom panel enables users to construct the molecules from different atoms. The core of the gestural detection algorithm of MolyPoly is to determine the users' inferred gesture from the position of their detected hands in successive Kinect frames.

Gestures were chosen based upon their ease of use and whether they felt natural. In addition, one of the limitations of the Kinect was the ability to track an individual hand across the crossover workspace (e.g. left hand reach over to the right-handed side of the body): therefore, we designed all interaction to have no crossover and therefore individual hands only operated on their side of the body. During the development and pilot test, we trialed a number of gestures, but some were found too hard to use by the pilot testers. For example, an alternative method of rotation was trialed by simply making the molecule face the direction of the cursor. However, it was not chosen due to lack of control and fine rotations. The eventual gestures were found natural, intuitive and easy to perform. The following paragraphs describe the available gestures in further detail;

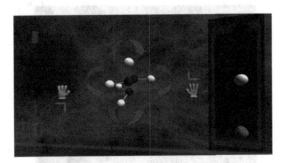

Fig. 6. Workspace with annotated arrows to indicate the directions in which the molecule structure can be rotated.

Rotation. Figure 6 illustrates how the molecular structure can be rotated 360 degrees on the x, y and z axis (note y is vertical axis and x is horizontal axis). The rotation is performed by moving the virtual hands (representing the user's hands) in different directions, as detailed below:

Left hand:

- Move hand down to rotate the structure downward around the x-axis.
- Move hand left to rotate the structure clockwise around the y-axis.

Right hand:

- Move hand up to rotate the structure upwards around the x-axis.
- Move hand right to rotate the structure counter clockwise around the y-axis.

Scale/Zoom. The detection of the zoom gesture was determined by moving both hands either inwards or outwards (in the horizontal axis as illustrated in the top-down view in Fig. 7). A velocity threshold value is used to avoid accidental and unintentional hand movement.

Fig. 7. (Top-down view) Performing the zoom in/out functions. Zooming out by moving both your hands from point 2 to 1 (inwards), while zooming in by moving in the reverse direction from point 1 to 2 (outwards).

Joining Atoms. Atoms can be joined and connected to the molecular structure by picking (selecting) them up and bringing them closer to the main structure. When the atom is in close proximity to another atom that has already been joined and therefore a part of the main molecule structure and still has available link(s), it will display a bond

Fig. 8. Close proximity to the structure displays the available bonds

showing the user that there is an available connection (gray tip in Fig. 8, near the right virtual hand) that can be connected to. The gray tip will change to a red tip when the user moves the selected atom very close to the tip and if the bond's tip stays red for three seconds, the held atom will be bonded (jointed) to the main molecular structure.

5 Experiment

To the authors' knowledge, there has been no research conducted explicitly exploring the effect of using an immersive 3D environment with gesture control on conceptual understanding of foundational chemistry. The aim of our pilot study is to determine whether or not the concept of the MolyPoly's 3D immersive gesture controlled interface can be utilized as an effective teaching tool within the field of chemistry by demonstrating that this technology has the potential to yield beneficial learning outcomes comparable to a traditional teaching method.

The experiment consisted of four sessions over two weeks. Seventeen (17) students from a first-year chemistry class at the University of Tasmania took part in the study. The participants were divided into two groups, MolyPoly and the traditional classroom group based on their scores in a pretest such that the two groups had similar ranges of students' prior subject knowledge. The traditional group of students remained with their normal chemistry classes and created molecular models using physical molecular model kits, whereas the MolyPoly group studied in the HIT Lab AU's VisionSpace, where they used the MolyPoly system to visualize and create molecular structures in conjunction with the printed handout of the lessons, which was used by both groups.

The pretest was conducted to ascertain prior knowledge in order to get a baseline for comparison with posttest after the lessons. The pretest and posttests were identical and consisted of basic chemistry knowledge-based questions, mainly on molecular structure. In the subsequent two weeks, four sessions with five molecular structures were taught to both groups separately using the two different pedagogies.

In the first session, participants selected for the MolyPoly group were shown how to interact with the application so that they had an opportunity to get acquainted with the system. Then, participants were asked to create a simple molecule structure, Methane, while participants in the traditional group were given a traditional lecture about the same structure. During the second session, all participants were required to create Propane, which is a more complex molecule structure. In the third session (second week), 1-Propanol was used in the third session, where the participants were instructed to complete the 1-Propanol structure and upon successful completion they were then able to view a completed 2-Propanol structure for comparison. In the (last) fourth session they were instructed to complete a Methoxymethane model. After all participants had completed the lesson, both groups undertook the posttest.

5.1 Pretest and Posttest Questions

Students responded to 15 questions on a pretest/posttest evaluation designed to elicit their conceptual understanding of foundational chemistry. These questions were categorized by level of difficulty (easy, moderate, difficult) and by level of required spatial knowledge (none, low, high). Each difficult level contained five questions. There were two questions with non-required spatial knowledge and both were in the easy difficulty level. There were four questions with low-level spatial knowledge with two in the easy difficulty level and two in the moderate difficulty level. The questions with requirement of high-level spatial knowledge contained nine questions that were equally divided across all levels of difficulties (three in each). In summary, the five easy questions required either high spatial knowledge or no spatial knowledge. The five moderate questions required some spatial knowledge (either low or high). Similarly, the five difficult questions required either low or high spatial knowledge.

Example questions are listed below (answers are highlighted in red and in the line immediately below the question line):

- [Easy, High Spatial Knowledge] Draw the full 3D structure for CH_4.

- [Easy, No Spatial Knowledge] The following Lewis Dot Diagram represents Propan-1-ol.

What type of hybridization does the oxygen atom have?
sp^3

- [Medium, Low Spatial Knowledge] Why would Propane have a lower boiling point compared to 1-Propanol?
  1-Propanol is able to easily form H-bonding and dispersion forces between molecules whereas Propane is only able to form dispersion forces between molecules.

- [Medium, High Spatial Knowledge] Draw the full 3D structure for propan-2-ol.

- [Difficult, High Spatial Knowledge] Put the following molecules in increasing order according to their boiling point. CH_3CH_3OH, $CH_3CH_2CH_2OH$, and $(CH_3)_2CH_2OH$.
  3, 2, 1

At the end of the MolyPoly group experiment, participants were asked to fill in a questionnaire to rate their experience of using the MolyPoly application. There were three main questions: (1) rotation gestures easy to use, (2) zoom gestures easy to use and (3) gestural manipulation improve understanding of the molecular structures. The answers were Likert-scale with 1 Strongly Disagree, 2 Disagree, 3 Neutral, 4 Agree, and 5 Strongly Agree.

5.2   Analysis Strategy

All tests were scored by a member of the research team, who was the subject domain expert and a regular lecturer for the unit. Each question was marked as either right or wrong. All correct responses were tallied to provide the dependent variables for inferential analyses between the two groups (traditional and MolyPoly) and the pretest and posttest. The reliability of this analysis was strengthened by the collaborative marking effort that provided for a moderation process to resolve any conflicting interpretations.

A 2 (group) × 2 (test) mixed design ANOVA, with an alpha level set at 0.05, was used to test for any significant differences between the summative scores of the 15 questions. This was followed by a series six 2 (group) × 2 (test) mixed design ANOVAs to test for significant differences of the summative scores for each of the question difficulty levels ((5) easy, (5) medium, (5) difficult) and level of spatial knowledge ((2) none, (4) low, (9) high) individually. Since several statistical tests were performed, a modified Bonferroni adjustment was applied to the overall critical alpha level before any significant differences were identified. Only statistically significant findings are reported below. All statistical analysis was conducted with SPSS version 22.0.

5.3   Results

There were no statistical differences between the two groups on the pretest evaluation. Both groups improved their foundational knowledge of chemistry over the experimental period. Our analysis indicated that both groups improved their pretest/posttest performance similarly across all question difficulties. There was a statistically significant improvement for the easy and moderate questions, however there was no statistically significant improvement for the difficult questions. There was a statistically significant improvement for the low and high spatial knowledge questions; however, there was no statistically significant improvement for the two questions that contained no spatial knowledge. In summary, one group did not perform better or worse compared to the other group for any of the pretest/posttest comparisons. The details of our significant findings are described in the following paragraph.

The overall analysis of the total 15 questions revealed a significant main effect for test, $F(1,16) = 110.03$, $p = 0.001$. Both groups significantly improved from pretest ($M = 5 \pm 1$) to posttest ($M = 10 \pm 2$), total score of 15. When categorized by level of difficulty, analysing the five easy questions showed there was a significant main effect for test, $F(1,16) = 95.31$, $p = 0.001$. Both groups significantly improved from pretest ($M = 2 \pm 1$) to posttest ($M = 4 \pm 1$). For the five moderate questions, there was a significant main effect for test, $F(1,16) = 32.82$, $p = 0.001$. Both groups significantly improved from pretest ($M = 1 \pm 1$) to posttest ($M = 3 \pm 1$); and for the five difficult questions, there was no significant findings for the improvements noted by both groups, pretest ($M = 2 \pm 1$) to posttest ($M = 2.5 \pm 1$).

For the four questions that required low spatial knowledge, there was a significant main effect for test, $F(1,16) = 22.94$, $p = 0.001$. Both groups significantly improved from pretest ($M = 2 \pm 1$) to posttest ($M = 3 \pm 1$). For the nine questions that require high spatial knowledge, there was also a significant main effect for test, $F(1,16) = 52.07$, $p = 0.001$. Both groups significantly improved from pretest ($M = 2 \pm 1$) to posttest ($M = 5 \pm 2$).

The time durations in which both groups were working with their respective 3D molecular forms (traditional group–molecular model kit in Fig. 1, MolyPoly group – 3D molecular models in Fig. 3) in each of the lessons were on the average 4.71 min per lesson per group class, across all five lessons, for the traditional classroom group and 2.94 min per lesson per person for the MolyPoly group. The small difference in these exposure times that individuals spent on 3D models were not significant in the context of this study as the small differences were expected due to non-identical teaching pedagogies of the two groups. For the rest of the time (50-min teaching session), the traditional group would go through (typical) passive learning (with and without the use of the model kit) on the printed out lesson materials and Q&A without the use of the model kits. Simpler for the MolyPoly group, students on their own spent the rest of the time learning from the printed out lesson materials, which are the same materials used in the classroom group. As opposed to a traditional group's teacher-driven teaching session, the MolyPoly group environment facilitated student-driven learning or self-direct learning.

The post experiment questionnaire revealed that participants felt more comfortable using Zoom gestures (with an average response of 4.14) than using Rotation gestures (with an average response of 3.14). Furthermore, participants generally agreed that the ability to manipulate the molecular structures in 3D (through MolyPoly) helped them improve their understanding of the structures (with an average response of 3.85).

6 Discussion

Our results showed that the 3D immersive virtual environment created by MolyPoly produced posttest score results comparable to posttest score results of the group with the traditional teaching method. This implied that the use of MolyPoly's 3D immersive gesture controlled interface was as effective as the use of the physical model kits in the spatial learning context. The learning outcomes of the MolyPoly group were results from a combination of using 3D immersive gesture controlled approach and self-direct active learning approach.

MolyPoly's interactive 3D immersive gesture controlled approach has built-in constraints and feedback that allow students to explore the mapping between 2D diagrams and 3D structures. Students were able to use gestures to control the viewpoint of the 3D Molecule structures; spin the molecule to the left, right, up, and down, and additionally zoom in and out of the structures. Combining the gestural control with 3D immersive capabilities created a learning environment that facilitated visuo-spatial reasoning and learning thorough the active learning and constructivist approach. Students were able to actively map what they learned from the lesson information sheets (2D diagram) to their spatial understanding of the 3D molecule structures through the process of constructing those molecules atom-by-atom in a 3D structural manner. Initially, the students had trouble performing the gestures, but they eventually became quite adept at using them effectively following the first lesson. However, as the post experiment questionnaire revealed, participants found rotation gestures more difficult to use than zoom gestures. The design of the rotation gestures required some cognitive load to operate the two-hands gestures properly; while the inwards and outwards of the zoom gestures were more natural. However, it is worth mentioned that despite some less than natural interaction of our gestures, participants still felt that their ability to manipulate the 3D molecular structures helped them improve their understanding of the molecules. The future work should include the redesign of the gesture interaction that would improve the naturalness of the interaction.

With MolyPoly's in-built lesson feedback, there was no need for a chemistry teacher. This allowed the students to create their own knowledge based on an engaging, meaningful experience. In a traditional teaching method (i.e. classroom), the chemistry teacher often directed the student attention to the "right answer" or informed them what was important to attend to. In the MolyPoly (i.e. self-direct), the students were forced to become active participants in constructing their own learning, allowing them to proceed at their own pace to forge their own conclusions and knowledge. This self-directed active learning style was also evident by the utilisation of the in-built lesson feedback which was used by students in the MolyPoly group on average of 1–3 times per lesson, whereas the traditional classroom group had virtually no direct questions from students during classes. This was most likely attributed to the capacity of MolyPoly to provide real-time, user-initiated augmented feedback that could be elicited as often as requested by each participant within this group. This functionality created by the immersive software environment allowed for more student-driven, individualized, problem-based learning experiences during the intervention period; whereas the traditional group relied on a typical teacher-driven learning experience that did not cater for individual rates of learner progress. Student-centered approaches to learning have often produced better-quality outcomes compared to traditional teacher-centered instructional approaches. For example McDonnell [35] found that chemistry students had a much more engaging learning experience when involved in a 'unique' project, as opposed to the traditional laboratory approach. In Chan's study [34], the participants found learning a dance lesson with a motion-controlled VR dance training system with real-time on-screen feedback much more interesting and highly engaging than simply learning from watching videos.

We believe that if our experiment was replicated across a larger group of students over a longer experiment period, the greater power elicited would demonstrate that the MolyPoly teaching environment would yield better learning outcomes compared to the more traditional teaching environment conducted for introductory level chemistry classes. Furthermore, it was the aim of the researchers to test the amount of learning gained through a 15 question posttest. It would be advantageous for future related research studies to incorporate a retention test so true learning using this technology can be measured.

Our results also implied that the MolyPoly teaching environment has the potential to be a viable teaching solution for remote delivery of knowledge subjects that requires not only spatial reasoning and learning but also self-directed active learning. Future research should explore the possibility of this potential by conducting a field study with real remote classrooms. The un-joining function should also be implemented in the future work so that students can undo the unwanted joining.

7 Conclusions

MolyPoly is a proof of concept of a basic 3D molecule construction simulator, specifically designed to help first-year university chemistry students learn organic chemistry. It was designed to utilize a mixed approach of 3D immersive gesture controlled and self-directed active learning to facilitate visuospatial learning of organic chemistry. The aim of the research was to determine through a pilot study whether MolyPoly could be used as an effective teaching tool for teaching organic chemistry. A pilot study was designed and conducted to compare the comparative learning outcomes between the group which learned with MolyPoly and the group which learned in a traditional classroom. The analyses of the pilot study provided strong evidence that students who learned using MolyPoly demonstrated a pretest-posttest score improvement of learning outcomes comparable to students that learned in traditional classroom.

Comparing the pretest/posttest over different difficulty levels of organic chemistry suggested that the students in the MolyPoly group performed on the tests just as well as the students in the traditional classroom group. Additionally, the two groups performed similarly on the questions, which required different levels (no, low, and high) of spatial knowledge. This is a further indication that the MolyPoly has potential as a teaching tool for the subjects that require spatial reasoning and learning such as organic chemistry.

MolyPoly's 3D immersive environment with gestural controls created the learning environment that promoted visuo-spatial reasoning by externalizing complex spatial relations and providing interactive gesture controlled tools for direct manipulation of spatial relations. Furthermore, MolyPoly's in-built lesson feedback facilitated the self-directed active learning, which promoted active students' engagement in constructing their own learning. This further improved the students' understanding of the subject materials.

Acknowledgements. We would like to acknowledge the support of a Provost Visiting Scholars Program at the University of Tasmania that enabled us to form an international research team. We also would like to acknowledge the contribution of an undergraduate final year project team, MolyMod, who in 2012 helped create the motivation and original conceptual prototype upon which this project was built and Jonathan O'Duffy of the HIT Lab AU for his help during the initial development of MolyPoly in 2013.

References

1. Ferk, V., Vrtacnik, M., Blejec, A., Gril, A.: Students' understanding of molecular structure representations. Int. J. Sci. Educ. **25**(10), 1227–1245 (2003)
2. Wu, H.K., Shah, P.: Exploring visuospatial thinking in chemistry learning. Sci. Educ. **88**, 465–492 (2004)
3. Barnea, N., Dori, Y.J.: Computerized molecular modeling as a tool to improve chemistry teaching. J. Chem. Inf. Comput. Sci. **36**(4), 629–636 (1996)
4. Copolo, C.E., Hounshell, P.B.: Using three-dimensional models to teach molecular structures in high school chemistry. J. Sci. Educ. Technol. **4**(4), 295–305 (1995)

5. Dori, Y.J., Barak, M.: Virtual and physical molecular modeling: Fostering model perception and spatial understanding. Educ. Technol. Soc. **4**(1), 61–74 (2001)

6. Lohman, D.F.: Spatial ability. In: Sternberg, R.J. (ed.) Encyclopedia of Intelligence, vol. 2, pp. 1000–1007. Macmillan, New York (1994)

7. Wai, J., Lubinski, D., Benbow, C.P.: Spatial ability for STEM domains: Aligning over 50 years of cumulative psychological knowledge solidifies its importance. J. Educ. Psychol. **101**(4), 817–835 (2009)

8. Rafi, A., Anuar, K., Samad, A., Hayati, M., Mahadzir, M.: Improving spatial ability using a Web-based Virtual Environment (WbVE). Autom. Constr. **14**(6), 707–715 (2005)

9. Hegarty, M., Waller, D.: A dissociation between mental rotation and perspective-taking spatial abilities. Intelligence **32**(2), 175–191 (2004)

10. Keehner, M., Lippa, Y., Montello, D.R., Tendick, F., Hegarty, M.: Learning a spatial skill for surgery: How the contributions of abilities change with practice. Appl. Cogn. Psychol. **20**, 487–503 (2006)

11. Fulmer, L., Fulmer, G.: Secondary Students' Visual-Spatial Ability Predicts Performance on the Visual–Spatial Electricity and Electromagnetism Test (VSEEMT). Sci. Educ. Rev. Lett. **2014**, 8–21 (2014)

12. Chen, Y.-C.: A study of comparing the use of augmented reality and physical models in the chemistry education. In: Proceedings of the 2006 ACM International Conference on Virtual Reality Continuum and Its Application, pp. 369–372 (2006)

13. Dunser, A, Steinbugl, K, Kaufmann, H., Gluck, J.: Virtual and augmented reality as spatial ability training tools. In: Proceedings of the Seventh ACMSIGCHI New Zealand Chapter's International Conference on Computer-Human Interaction, pp. 125–132 (2006)

14. Feng, J., Spence, I.: Playing an action video game reduces gender differences in spatial cognition. Psychol. Sci. **18**(10), 850–855 (2007)

15. Mohler, J.L.: Using interactive multimedia technologies to improve student understanding of spatially-dependent engineering concepts. In: Proceedings of the International GraphiCon Conference on Computer Geometry and Graphics, pp. 292–300 (2001)

16. Rafi, A., Samsudin, K.A., Ismail, A.: On improving spatial ability through computer-mediated engineering drawing instruction. Educ. Technol. Soc. **9**(3), 149–159 (2006)

17. Kosslyn, S.M.: Image and Mind. Harvard University Press, Cambridge (1980)

18. Shepard, R.N., Cooper, L.: Mental Images and Their Transformation. MIT Press, Cambridge (1982)

19. Sweller, J.: Cognitive load theory, learning difficulty, and instructional design. Learn. Instr. **4**(4), 295–312 (1994)

20. Tversky, B.: Visuospatial reasoning. In: Holyoak, K., Morrison, R. (eds.) The Cambridge Handbook of Thinking and Reasoning, pp. 209–240. Cambridge University Press, New York (2005)

21. Rauscher, F.H., Krauss, R.M., Chen, Y.: Gesture, speech, and lexical access: The role of lexical movements in speech production. Psychol. Sci. **7**, 226–231 (1996)

22. Iverson, J., Goldin-Meadow, S.: What's communication got to do with it? Gesture in children blind from birth. Dev. Psychol. **33**, 453–467 (1997)

23. Smetana, L., Bell, R.: Computer simulations to support science instruction and learning: A critical review of the literature. Int. J. Sci. Educ. **34**(9), 1337–1370 (2012)

24. Harrison, A.G., Treagust, D.F.: Modelling in science lessons: Are there better ways to learn with models? School Sci. Math. **98**(8), 420–429 (1998)

25. Sauer, C., Hastings, W., Okamura, A.: Virtual environment for exploring atomic bonding. In: Eurohaptics, pp. 5–7 (2004)

26. Erenay, O., Hashemipour, M.: Virtual reality in engineering education: A CIM case study. Turkish Online J. Educ. Technol. **2**(2), 51–56 (2003)

27. Mcgrath, D., Wegener, M., Mcintyre, T.J., Savage, C., Williamson, M.: Student experiences of virtual reality - a case study in learning special relativity. Am. J. Phys. **78**(8), 862–868 (2009)
28. Noeth, R.J., Volkov, B.B.: Evaluating the Effectiveness of Technology in Our Schools. ACT Policy Report (2004). http://www.act.org/research/policy/indeex.html
29. Grantcharov, T., Kristiansen, V., Bendix, J., Bardram, L., Rosenberg, J., Funch-Jensen, P.: Randomized clinical trial of virtual reality simulation for laparoscopic skills training. Br. J. Surg. **91**(2), 146–150 (2004)
30. Alexander, A., Brunyé, T., Sidman, J., Weil, S.: From gaming to training: A review of studies on fidelity, immersion, presence, and buy-in and their effects on transfer in pc-based simulations and games. In: DARWARS Training Impact Group, pp. 1–14 (2005)
31. Watson, W.R., Mong, C.J., Harris, C.A.: A case study of the in-class use of a video game for teaching high school history. Comput. Educ. **56**(2), 466–474 (2011)
32. Dalgarno, B., Lee, M.J.W., Carlson, L., Gregory, S., Tynan, B.: An Australian and New Zealand scoping study on the use of 3D immersive virtual worlds in higher education. Australasian J. Educ. Technol. **27**(1), 1–15 (2011)
33. O'Rourke, M.: Using immersive 3D computer games to help engage learners and deliver skill sets. In: 16th Australian Vocational Education and Training Research Association Conference (2013)
34. Chan, J., Leung, H., Tang, J., Komura, T.: A virtual reality dance training system using motion capture technology. IEEE Trans. Learn. Technol. **4**(2), 187–195 (2011)
35. McDonnell, C., O'Connor, C., Seery, M.K.: Developing practical chemistry skills by means of student-driven problem based learning mini-projects. Chem. Educ. Res. Pract. **8**(2), 130–139 (2007)

Sustaining Cognitive Diversity in Collaborative Learning Through Shared Spatially Separated Virtual Workspaces on Mobile Devices

Mark Reilly[1], Haifeng Shen[1(✉)], Paul Calder[1], and Henry Duh[2]

[1] School of Computer Science, Engineering and Mathematics,
Flinders University, Adelaide, Australia
{m.reilly,haifeng.shen,paul.calder}@flinders.edu.au
[2] HIT Lab Australia, University of Tasmania, Launceston, Australia
henry.duh@utas.edu.au

Abstract. Student disengagement in lectures is a global issue in higher education. Our approach is to apply a student-centred collaborative learning pedagogy into the lecture environment through a mobile real-time collaborative note-taking application, which allows a small self-selecting group of students to proactively keep each other engaged without disrupting others in the lecture venue or requiring changes to the existing pedagogy. We first present the application interface, which enables students to follow principles identified as good practice in undergraduate education while still allowing for the individual to contribute and interact with regard to their own ability and preferred learning style. The interface provides an individual virtual workspace to each group member, which is shared with and may be viewed and edited by other members in the session using their own mobile devices. A pivotal design goal is to accommodate students' diverse cognitive abilities by spatially separating their workspaces and allowing each individual to choose the most suitable way to interact with their peers' workspaces. We then discuss the results of experiments that compared between individual and collaborative note-taking and between using shared common workspaces and spatially separated individual virtual workspaces. The results show that students are more engaged in the lecture with collaborative than individual note-taking and more satisfied with sharing spatially separated workspaces than a common workspace.

Keywords: Engagement · Computer supported collaborative learning · Mobile learning · Multi-user interface · Workspace

1 Introduction

A lecture is still the primary teaching and learning paradigm in most universities, and is likely to remain so in many years to come. The most important benefit of the lecture to the university is its cost-effectiveness as a model of delivery to an audience ranging from tens to hundreds, and perhaps even thousands. It is where the uniform delivery of information, not found in the standard texts, and also the possibility to motivate and inspire students exist [3]. However, to many, a lecture is also seen as a sub-optimal learning environment with issues ranging from the shape of the venue promoting the

© Springer International Publishing Switzerland 2015
T. Wyeld et al. (Eds.): OzCHI 2013, LNCS 8433, pp. 171–193, 2015.
DOI: 10.1007/978-3-319-16940-8_9

audience as spectators rather than participants through to the sense of isolation many students report, from being a lone voiceless participant in a room of many, and up to hundreds, which are the major cause of student disengagement.

Our primary objective is to encourage students to proactively engage in a lecture by motivating them to actively take notes during the class, an effective learning technique that aids effective learning by fostering encoding, articulation and rehearsal [3]. The practice of taking notes during a lecture is long established as a means of capturing knowledge and understanding [8, 12], as is the benefit of revising those notes at a later time [12], and of discussing those notes with at least one other person [3]. While acknowledging that the absence of taking notes does not necessarily indicate a lack of engagement, we use the note-taking practice as a proxy for educationally meaningful activity within a lecture session. Our approach is to apply a student-centred collaborative learning pedagogy into the lecture environment through a mobile real-time collaborative note-taking application called GroupNotes. GroupNotes aims at improving the engagement of students during lectures by providing a communication platform on the devices such as tablets and smartphones to both enable and encourage student interaction [18]. This interaction reflects the social aspect common to learning, which is frequently absent from the didactic lectures that still exist today [9]. It allows a small self-selecting group of students to interact using their own devices, and motivate, assist, and monitor each other in order to actively learn and keep themselves engaged during the lecture, essentially transforming the sage-on-the-stage amphitheatre [7] into a student-centred collaborative classroom [27]. Groups will structure themselves according to their shared aims, and with respect to the knowledge, skills and abilities, of their group members.

The rationale behind the design of the application interface is to stimulate student engagement from a student-to-student perspective by incorporating the well-known Seven Principles for Good Practice in Undergraduate Education [4], which provide a roadmap of what works in undergraduate education. While the implementation of these principles has mostly been vertical, that is, for the lecturer to proactively interact and engage students, we are interested in a horizontal implementation, emphasising real-time interaction between small self-selecting groups of students in a manner that does not disrupt the lecturer, others in the vicinity, or others in the group. In order for the interaction to influence educationally meaningful behaviour, we provide each group member with real-time access to all written work of each other member and use peer pressure within these groups to encourage/enforce adoption of the seven principles from a student-to-student perspective.

An important inspiration for our research was Livenotes [11], which enabled small groups of students to jointly annotate lecture slides in real time using tablet PCs connected via wireless networks in a way that only required minimal institutional or pedagogical change and was independent of the size of the class in the lecture venue. However, its multi-user interface was a shared whiteboard that did not support structured notes. The single-document interface did not allow the individual student to concentrate solely on their own task, or flexibly choose their preferred manner of working with their peers. Therefore, a pivotal design goal is to accommodate students' diverse cognitive abilities by spatially separating their workspaces and allowing each individual to choose the most suitable way to interact with their peers' workspaces.

Test results show that students are more engaged in the lecture with collaborative than individual note-taking [20] and more satisfied with sharing spatially separated work-spaces than a common workspace [21].

The rest of the paper is organised as follows. The next section describes work related to our approach. We then present the application interface, followed by the design principles. After that, we present the tests and discuss the results. Finally we conclude the article with a summary of major contributions and future work.

2 Related Work

Common to the majority of lectures is a didactic style of presentation where the lecturer delivers content to the students in a one-way flow of information. Questions to the lecturer are generally not encouraged as they would interrupt the information flow and may possibly drive the lecture behind schedule, despite instant feedback being of most benefit to the students doing the asking, and presumably others in the audience as well. The second avenue to obtaining quick feedback during the lecture is discussions between students however there is a real or implied prohibition on this as it may disrupt the lecturer and students in the vicinity. Students can only get the full benefits of the lecture if they are fully engaged. However, the traditional didactic style of lecture - with unidirectional flows and little or no interaction between lecturer and students, and among students themselves - is a significant cause of student disengagement [9].

Much attention has been paid recently to re-design lectures to incorporate vertical lecturer-to-student interaction [18], such as short quizzes, instant questions, or real-time voting, supported by electronic devices like clickers [15] or mobile phones [6, 11, 16]. Empirical evidence as to how student engagement can be improved as well as to the effectiveness of such strategies in increasing student learning outcomes is yet to be proven [17], but the use of mobile devices does increase attendance in courses [5]. The technical sophistication and portability of mobile devices remove the major obstacles to providing students viable technological solutions to their learning needs. Saville-Smith suggested that handheld computers can assist students' learning in a number of ways, including motivation and supporting both independent and collaborative learning [22]. Milrad identified useful features that mobile technologies provide for education in terms of social interactivity, individuality, connectivity and portability [16].

The work of Notetaker [28] suggested that the ideal electronic device for taking notes required a keyboard for the speed improvement over handwriting, a pen to enable faster and superior drawing than from a mouse while still needing the mouse for exact positioning of text and drawing within the page. While larger devices such as a tablet PC may provide a more usable learning environment [11, 24], not all students are able or willing to purchase such a device, bring it to university with them, and then actually use it in the lecture environment because voluntary self-ownership and familiarity with their own device are the key to removing the major learning curves associated with new technology. Conversely, there will be no additional learning or monetary costs to use students' already-owned mobile devices in the lecture environments [14]. The touch-screen interface, possibly with a stylus, allows for exact positioning and ease of drawing and text entry. Research on using smartphones for teaching and learning is emerging in recent years primarily due to increased self-ownership and the familiarity with their own devices. Combined with the technical sophistication in computation power, screen size, storage, and easy-to-use interface, as well as portability and wireless connectivity, they remove the major obstacles to providing students viable technical solutions to their learning needs as evidenced with the recent introduction of Hotseat [1] and MLI [10].

3 Design of the Application Interface

GroupNotes aims to provide a mobile platform that introduces the social interaction common to learning into a lecture environment. The aim of the interface design is to provide the individual student with the flexibility to control the level of interaction they are exposed to, thereby allowing them to work in the manner that best suits themselves while at the same time contributing to the engagement and learning outcomes of the other members they have selected to work with. Individual users can see all content generated by other members in their group in real time and in a form that enables immediate identification of the author through a combination of unique avatar/colour/ username that persists throughout the entire session and is consistent on all devices. GroupNotes does not enforce the same interface onto all group members' devices, enabling individuals to choose when and where to access the content according to personal characteristics such as cognitive ability, ease of distraction and so on.

The structure of Notes mimics that of the lecture. Individuals provide comments in either the drawing or text workspace of a Note, which corresponds to the lecture slide. For example, comments on the n-th page of a Note relates to content of the n-th lecture slide. Each individual owns a set of Notes, which are shared with typically 2–4 other members. The size of the group is generally limited by the cognitive difficulty of keeping up with both the lecturer and their peers [19]. A session of n students shares $2n + 1$ multi-page workspaces, including the lecture slides and n Notes each consisting of a text and a drawing workspace. The spatially separated yet sharable virtual workspaces are the key design innovation that provides students the maximal flexibility to work with their peers by respecting their diverse cognitive abilities.

Each individual is allowed to edit their peers' Notes, but the one who created a Note owns the entire Notes, including all comments inside it, regardless of who made them. This means that when viewing the content of a Note, one may see multiple colours in text and in drawings; this makes sense because the reading of the content as a modification or addition to an individual's Notes will generally only make sense when you see them in context. All content generated either in their own or others' Notes occurs in the local device and at the same time is transferred to a server in the cloud where synchronisation takes place before the content is delivered to other members' devices [23].

As shown in Fig. 1, the interface comprises three separate tabbed sections: Notes, Lecture Slides, and Community Notes. The Notes tab has an Editing Pane (upper) and a Viewing Pane (lower) divided by a splitter. The Note, when visible in the Editing Pane, contains a vertical identification bar on the left, including the corresponding slide number, the unique avatar/colour/username associated with all content generated by the

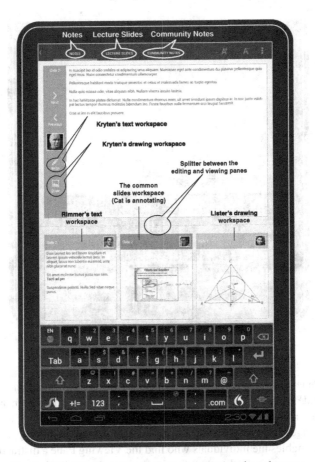

Fig. 1. Multiple shared spatially separated virtual workspaces

owner of the Note, and navigation arrows to traverse backwards or forwards through the available pages and switch between the text/drawing workspaces.

Content generation is only possible when a Note appears in the Editing Pane and is limited to a single Note at a time on a device. The user who created the Note would generate content to which other group members may further add, either through additional comments or emphases on the originator's content by such means as highlighting or underlining the text or drawing using their own unique colour. They may also choose to disagree with the originator's content, if it is text by striking through text, or if it is drawing, by drawing over the original using their own unique colour. The use of unique colour to identify the contribution of each member of the group allows each user to determine how much weight another user gives to that content in relation to their own knowledge validation or construction.

The Viewing Pane shows the most recently edited pages of group members' Notes or the slide that is currently being annotated. In Fig. 1, pages in Rimmer's text workspace in the Viewing Pane are showing Slide 2, the same as Kryten's text workspace in

Fig. 2. Minimised viewing pane

the Editing Pane and Lister's drawing workspace in the Viewing Pane, while Cat is annotating Slide 2 in the common slides workspace. Content within an individual workspace shown in the Viewing Pane is viewable (scroll up and down to view the full content of a page) but not editable. The splitter dividing the Editing Pane from the Viewing Pane enables the individuals who find the Viewing Pane a distraction to reduce its on-screen size until only the identification bars of the available group members are visible, as shown in Fig. 2.

Access to other than the most recent content generated by a group is available through the read-only Community Notes tab as shown in Fig. 3 or by dragging the Notes of that group member into the Editing Pane for viewing or even editing. As soon as a user has a Note in the Editing Pane, they are able to navigate backwards and forwards through the entire history of that Notes' text/drawing workspaces and generate their own content into those pages wherever they choose.

The second tabbed area is for the Lecture Slides. The content in this workspace differs from a Note in that it is owned by the group and any member can annotate any slide as required. The inclusion of the lecturer's slides is important for two reasons. Firstly it is these slides which will be incorporated into the Community Notes and must be accurate at the time the lecture is delivered, with any modifications required by the lecturer, such as correcting a mathematical formula, an incorrect date or some other important point or by the group members, such as a comment tightly coupled with the slide context. The second reason is to provide greater flexibility for the users in their preferred manner of taking notes with the option of annotating the lecturer's slides directly.

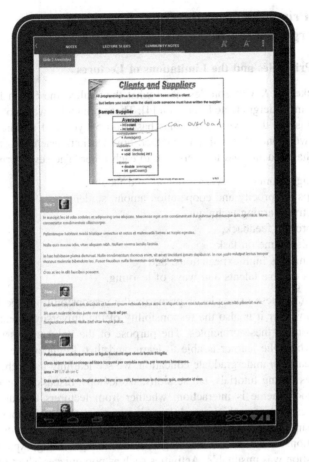

Fig. 3. The community notes with individual notes and annotated slides

The third tabbed area is Community Notes, which serves a dual purpose: as a running commentary of the notes and slide annotations generated during a session and as a complete record of all generated content for the session, which can then be reviewed at a later stage. During a session, all content generated in the pages of a Note is delivered in real time to its own section of the page in the Community Notes. This allows a user to view the entirety of an individual's thoughts along with any additional comments made by others related to a specific lecture slide's content in a single view, and then do the same with each other group member's generated content for the same slide. The second stage of the Community Notes is delivered when the last entry to the Notes has been completed by all group members. At this point, a single PDF document is created by the server, where each page contains the (annotated) lecture slide followed by the corresponding page of the note from each group member, as shown in Fig. 3.

4 Design Principles from the Good Practice in Undergraduate Education

4.1 Seven Principles and the Limitations of Lectures

In 1987, Chickering & Gamson described what they called the Seven Principles for Good Practice in Undergraduate Education [4]. These principles were not intended as a silver bullet, instead they were guidelines based on 50 years of research identifying how students work and play with one another, how teachers teach and students learn, and the way that students and teachers talk to each other. The seven principles are:

- P_1: encourages contact between students and faculty,
- P_2: develops reciprocity and cooperation among students,
- P_3: uses active learning techniques,
- P_4: gives prompt feedback,
- P_5: emphasises time on task,
- P_6: communicates high expectations, and
- P_7: respects diverse talents and ways of learning.

Implementation of these principles is primarily the responsibility of the lecturers and the students, however it is also the responsibility of the university to enable the conditions that promote these principles. The purpose of the university is to provide an environment where the learner is able to learn, and then to encourage the learner to learn how to learn. In undergraduate education, we see learning using these principles on a daily basis during tutorials, lab sessions, workshops, discussion groups, forums, etc. The common theme is interaction, whether from lecturers to students or from students to other students, which involves active participation with others.

Lectures can implement these principles through various methods such as by allowing questions at the correct teaching moment and providing explanations, even where the question was unsuitable. Activities such as pop quizzes, short relevant video clips, even a few minutes time out to allow for contemplation, catch-up on notes, questioning of peers all encourage participation. Unfortunately this interactive type of lecture is not universal; the traditional didactic lecture with its one-way information flow still is common, and many students do not like it, either by not attending or participating in activities indicating their displeasure such as using Facebook or Twitter, surfing the internet, playing games or even sleeping.

The lecturer controls the lecture environment and sets the tone for interaction. If the tone is that of the "sage on the stage" where the information is presented as though a performance, the audience are not active participants in their own knowledge construction; instead they are passive observers. This type of setting precludes even interaction between audience members. In much the same way as a cinema goer complies with the expected behaviour at the movies and does not talk while the show is on, students do not talk while the lecture is on. The important point here is talk; there is a real or implied prohibition on disrupting or distracting others in the audience from the lecture delivery and until recently there has not been an alternative means of communication to voice. Recent mobile devices are suitable for fast, silent, written interaction with multiple collaborators during a lecture. Moreover, these devices are already popular, and considered very desirable for daily activities, such as Internet access, Facebook, Twitter, reading, listening to music, and watching movies.

4.2 Implementation of the Seven Principles from the Student-to-Student Perspective

Our hypothesis is that these devices, when equipped with custom-developed software can reproduce the interactivity missing from didactic lectures. GroupNotes implements these principles mainly from the perspective of students because learning involves the reconciliation of multiple perspectives, including that of the peers of the learner [13]. The sharing of notes with other learners may encourage students to improve their notes, to critically examine their own understanding, and to co-construct a shared understanding with other learners [26]. In the instance of a didactic lecture, the lecture theatre environment is explicitly constraining the ability of the individual learner to access the perspective of their peers at the appropriate 'teaching moment' when it would provide the greatest benefit. GroupNotes provides a student-to-student implementation of principles $P_1 - P_7$ in a manner that transforms the passive audience into active participants in their own knowledge construction without disrupting either the lecturer or the other students in the vicinity.

While not all groups, or all members in any group, will be motivated by every principle, we expect that at least several will be a catalyst to increase their engagement with the lecture. Examples of the type of interaction expected is cooperation to ensure complete coverage of the lecture notes (P_2) and active learning through the examples provided by other group members which illustrate their knowledge (P_3). The availability of these user-generated notes in real time provides prompt feedback to the user through validating or challenging the knowledge viewed in others notes (P_4). The knowledge on the part of the individual that their notes are immediately available to others in the group ensures that time is not spent on activities not related to the aims of the group (P_5). The group has self-selected their members and the proximity to the work of others promotes working to the level required to achieve the group's aim (P_6). The respect for diverse talents and ways of learning (P_7) is promoted by allowing for the group to cooperate/collaborate in any manner they choose as long as the members are contributing in an agreed fashion to the aims of the group.

First, P_2 encourages reciprocity and cooperation among students. GroupNotes enables students to form a group based on whatever criteria they choose without compromising the ability of any one individual to work in a manner that suits their own learning style. This self-selected group can then cooperate in a manner which best suits their shared goal of knowledge acquisition by allowing text to be written, drawings to be made, and slides to be annotated by these individuals, and with results available to all group members in real time. Figure 1 shows the device owner is editing text in their own note while simultaneously viewing the most recent editing activities of the other three group members in their workspaces: text on a note, drawing on a note, and annotation of a slide. The activities being carried out may be the result of cooperation with all members doing their own tasks or the result of a collaborative effort aiming to provide the best outcomes for the group by designating certain tasks to those best equipped to handle them, in this case two members taking text notes, another annotating the slides, and the final one doing all the drawings. Rather than the lecturer directing and encouraging cooperation and reciprocity, it is the self-selected members of the group who are organising their own group to meet their own goals in a manner that best suits them in achieving their goal.

Second, P_3 encourages using active learning techniques, which means students must talk and write about what they are learning, relate it to past experiences and apply it to their daily lives. The act of taking notes by a student to illustrate the content being delivered is a means of relating this new content to their existing knowledge and experiences [8, 12]. It is established as an important part of the process of codifying lecture content into knowledge that can be retrieved later. GroupNotes provides a platform for not only taking notes as quickly as traditional pen and paper but also for viewing other group members' notes illustrating their own knowledge and experiences regarding the same content. Providing this in real time, with the individual able to access them at the appropriate 'teaching moment' during the lecture enables the student to also query others in their group when they are unclear on a certain aspect of the content or to explain their own knowledge to others as required. The ability to view and discuss content of which they are uncertain, or disagree with, at the time of their choice, either during or after the lecture, brings multiple perspectives to what is traditionally an individual endeavour.

Third, P_4 advises giving prompt feedback, including knowing what they do and do not know, as well as identifying opportunities for improvement. The ability to view the content generated by the other members in their group as it is created provides immediate feedback in the form of validation of their own knowledge if the other members agree with them. Otherwise they have the opportunity to investigate further and create new knowledge through a negotiation process that may involve just reading the content generated by their peers and accepting it into their own world view or involve asking questions. They may choose to ask questions in their own note or in the note of another group member who they consider the most likely to be able to provide the answer, while it may be another group member who provides the right explanation with the right example that allows them to consolidate this knowledge after further review. Their ability to assist their other group members when they require assistance in understanding something and in being engaged in the lecture will highlight to themselves where they have opportunities to improve their knowledge. If they can quickly provide sufficient knowledge, an example or experience, to allow a fellow group member to continue being engaged in the lecture, and not give up due to the content no longer making sense, then they are identifying if they need to improve in those areas.

Fourth, P_5 emphasises time on task, that is learning to use one's time well. The major factor responsible for a lack of engagement during lectures is the absence of interaction. However with GroupNotes providing the ability to gain insight into another group member's understanding during the lecture, this interaction is returned. While this may not provide the impetus to remain on task on all occasions, the principle agent paradigm states that a worker will withdraw effort from a task whenever the supervisory agent is removed [2]. In GroupNotes, the supervision is always present in the form of each other group member, therefore negating the likelihood of reducing effort on the task at hand and performing unrelated tasks. The dynamics of the group, since it is self-selected with the purpose of attaining a specific goal, will encourage time on task through implicit motivation to perform for the people they have chosen to work with.

Fifth, P_6 promotes high expectations. The theory of bounded rationality implies that an individual does not always aim for an optimum solution; they aim for a solution that is the best they can achieve with the resources they have and are willing or able to contribute [25]. It is the same force that will apply to maintaining high standards due to the high expectations agreed to with others in a group. They have selected a group of individuals they want to work with for the shared goal of acquiring new knowledge to a predetermined high standard and they must strive to meet the commitment they have agreed to regardless of their attitude on the day.

Last, P_7 respects diverse talents and ways of learning, which is the main reason for implementing this type of technology into the lecture environment as the belief that one-size-fits-all is appropriate appears illogical given the increasing diversity of the student population. In these circumstances where there is a lack of interaction and with a real or implied prohibition on talking, the only option available is to provide a solution which does not affect or require any change from the lecturer themselves, either in preparation or behaviour. GroupNotes provides a platform to allow for social interaction to occur quickly and silently without providing any disruption to those not involved.

A particular member may choose to take verbatim notes while others only choose to capture the important parts; or only reads and modifies the notes of another in the group in real time, or listens and only makes a note where the lecturer places particular emphasis during a session. It also provides the flexibility to allow the individual to control their own interface. The individual user, based on their own learning style, preferences and abilities to cope with multiple information sources, will determine how many of these sources they access at any given moment, and do so by knowing they are not impacting others in their group. While their device will receive all the content generated by the other members in their group in real time, it is the individual who decides when and how to view it. The user can choose to minimise all screens in their Viewing Pane so that only the identification bars are visible, such as Fig. 2. The user can choose not to look at any other member's Notes at all; they choose to participate just to provide their content to others. The user may choose to keep the Viewing Pane minimised at all times other than when they are having difficulty with the content of a particular slide and then raise the Viewing Pane to view the current screens of their other group members, such as Fig. 1 or they may choose to refer to the Community Notes, such as Fig. 3.

5 Tests and Results

5.1 Hypotheses

To test our ideas, we designed experiments to determine whether increasing engagement from a student-to-student perspective is a viable solution without requiring the lecturer to change their pedagogy, and to verify that the peer interaction does not disrupt either the lecturer or any other students in the venue, including within the same group. We video-recorded four separate lectures on a topic with the intention of producing the type of non-interactive lecture that causes students to not engage [9] for the purpose of testing three hypotheses:

- H_1: students are more engaged in the lecture with collaborative than with individual note-taking,
- H_2: students are more satisfied with sharing spatially separated workspaces than with sharing a common workspace, and
- H_3: students are more engaged outside the lecture with collaborative than with individual note-taking.

Hypothesis H_2 is decoupled to six sub-hypotheses, including:

- H_{21}: sharing spatially separated workspaces keeps students more engaged than sharing a common workspace,
- H_{22}: sharing spatially separated workspaces is a better implementation of the *Seven Principles* than sharing a common workspace,
- H_{23}: sharing spatially separated workspaces is more helpful than sharing a common workspace,
- H_{24}: sharing spatially separated workspaces is less distracting than sharing a common workspace,
- H_{25}: sharing spatially separated workspaces is more beneficial than sharing a common workspace, and
- H_{26}: sharing spatially separated workspaces is more likely to persuade students to come to non-interactive lectures than sharing a common workspace.

Hypothesis H_3 is decoupled to three sub-hypotheses, which are:

- H_{31}: more students prepare for the lecture beforehand with collaborative than with individual note-taking,
- H_{32}: more students revise notes after the lecture with collaborative than with individual note-taking, and
- H_{33}: more students discuss notes with at least one other person after the lecture with collaborative than with individual note-taking.

5.2   Methology

Four peer interaction methods $M_1 - M_4$ were tested among which M_1 did not allow for peer interaction, M_2 and M_3 supported peer interaction through a shared common workspace, and M_4 supported peer interaction through multiple shared spatially separated virtual workspaces.

- M_1: no peer interaction.
- M_2: audible peer interaction with a shared common tangible workspace, where the note-taking occurred on large sheets of paper with each member using a different coloured marker pen.

- M_3: silent peer interaction with a shared common virtual workspace provided by *Google Docs* on Nexus 7 tablets, where members of a group took notes in the same document and everyone was exposed to all members' notes and permitted to edit anywhere in the common workspace.
- M_4: silent peer interaction with multiple shared spatially separated individual virtual workspaces provided by *GroupNotes* on Nexus 7 tablets, where each member had their own workspaces with the option to view or hide, and edit all other workspaces according to their own requirements and without forcing that view on other members in the group.

A total of 32 university students across multiple disciplines participated in four testing sessions T_1 to T_4, where the four methods were tested respectively in the order from M_1 to M_4. The students mainly arrived as pre-organised groups of 2, 3 or 4 with only 6 of the 32 participants asking to be placed into a suitable group. The testing was not part of any enrolled class workload of the students. Each group performed all tests isolated from any other group to ensure a uniform environment and all groups were tested in the same order from T_1 to T_4.

We distributed a pre-questionnaire to students before conducting the four test sessions in order to capture some demographic details as well as data related to attitudes to attending lectures and learning methods used in lectures. We also arranged a post-questionnaire after the four tests to revisit the questions related to their attitude towards attending and engaging with the lecture content before, during and after the actual lecture, assuming the availability of the full-fledged GroupNotes. The four testing sessions each consisted of the participants viewing a pre-recorded lecture (one for each test) and then completing a questionnaire to capture their opinions on how well that particular method adhered to principles $P_2 - P_7$ as well as how engaged this level of adherence made them feel. Following the questionnaire was a quiz examining the lecture content whose main purpose was to persuade participants to treat the exercise

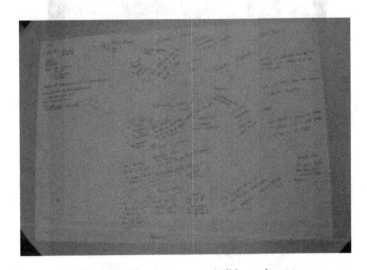

Fig. 4. A shared common tangible workspace

seriously. An additional purpose to the quiz in the long term is to assist with the design of tests that would measure the effectiveness of individual learning outcomes should the effectiveness of GroupNotes in engaging students prove to be significant.

M_1 was designed to set a baseline for each individual student about their normal practice when attending a non-interactive lecture, while $M_2 - M_4$ introduced groups and allowed interaction between members with the interaction method being the independent variable in the three experiments. M_2, as shown in Fig. 4, allowed peer interaction through a shared common tangible workspace, where the note-taking occurred on large sheets of paper (102 cm × 76 cm) with each member using a different coloured marker pen and audible communication among the members. We recognise this is not a viable option for lectures as the noise level would be too distracting and the desk space for a communal note area would require require a redesign of lecture venues. However, this method provided participants with the social environment in a lecture using our most natural method of social interaction, our voice, as well as introducing the communal note area. This method was also where participants were encouraged to develop some strategies that allowed them to maximise the groups and their own aims, regarding the type of note coverage they and the group wanted.

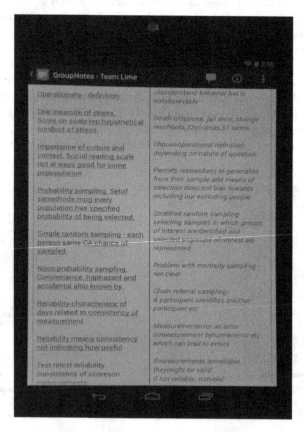

Fig. 5. A shared common virtual workspace

M_3 introduced mobile devices to the testing and allowed peer interaction through a common virtual workspace shared across Nexus 7 tablets (one for each group member), as shown in Fig. 5. The participants again watched a pre-recorded lecture and took notes according to what was most beneficial to the group without negatively impacting on their own performance. The major difference here is that audible communication was prohibited, requiring all communication, including coordination, to occur in writing, which makes it a viable solution for a lecture environment. The tablets were set up with a template in *Google Docs*, as shown in Fig. 5, allowing members of a group to take notes in the same document, where everyone was exposed to all members' notes and permitted to edit anywhere in the shared workspace.

M_4 differed from M_3 in that peer interaction was facilitated through multiple shared spatially separated individual virtual workspaces provided by GroupNotes as shown in Fig. 6 rather than a common workspace that everyone is confined to as provided by Google Docs. Each individual member had their own workspace with the option to view or edit all other workspaces according to their own requirements and without forcing that view on others in the group. It not only eliminated the distraction brought by others' note-taking activities but also offered the much needed flexibility for each individual to determine their own view of the content without affecting any other group member's view.

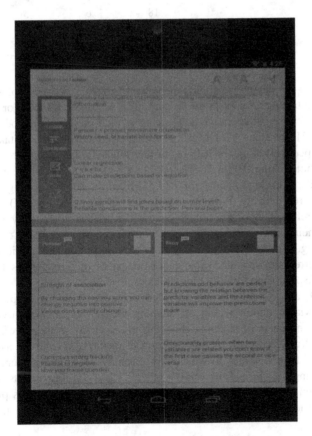

Fig. 6. Multiple shared spatially separated virtual workspaces

All groups performed tests in the order from T_1 to T_4, where a distinct set of variables were manipulated. T_1 involved no collaboration (M_1) while the remaining three tests did. T_2 and T_3 involved the use of common shared workspaces with T_2 having both audible and a shared tangible workspace (M_2), meaning there was no escaping from the input of others. T_3 removed the audible input from the testing but still provided a common shared workspace through a virtual one (M_3) on a mobile device. T_4 removed the common shared workspace, either audible or text-based, replacing them with multiple, virtual, spatially separated workspaces (M_4) which can be shared but did not force the immediate viewing of input from other group members.

5.3   Results

Preferred Note-Taking Method. Table 1 ($N = 32$) shows the choice of participants' preferred note-taking method after each of the four tests.

Table 1. Choice of preferred note-taking method after each test

Preferred method	After T_1	After T_2	After T_3	After T_4
M_1	–	14	7	5
M_2	–	18	6	3
M_3	–	–	19	0
M_4	–	–	–	24

Figures 7(a)–(c) depict the choice of preferred method between collaborative and individual note-takting after each test. In particular, 18 (or 56.2 %), 25 (or 78.1 %), and 27 (or 84.4 %) chose collaborative note-taking as their preferred method, while 14 (or 43.8 %), 7 (or 21.9 %), and 5 (or 15.6 %) still preferred individual note-taking, after T_2, T_3, and T_4 respectively. Figures 7(d)–(e) further illustrate that among those who chose collaborative note-taking, 19 (or 76 %) and 24 (or 88.9 %) preferred the quiet & virtual approach, while 6 (or 24 %) and 3 (or 11.1 %) liked the audible & tangible approach, after T_3 and T_4 respectively. With the introduction of M_4 in T_4, Fig. 7(f) shows that 24 (or 88.9 %) preferred sharing spatially separated workspaces and only 3 (or 11.1 %) stayed with their choice of sharing a common workspace. It is very encouraging to notice from Fig. 7(g) that although it is only a proof-of-concept application, all participants preferred the spatially separated workspaces of GroupNotes to the shared common workspace supported by Google Docs.

Method M_2 allowed talking and provided a shared tangible workspace. In reality, this is not a viable option within an ongoing lecture, which may account for the relatively high numbers of students choosing to select M_1 as their preferred method after T_2 (Fig. 7(a)). However, the introduction of methods M_3 and M_4 as more viable options opened their horizons to what is increasingly possible and therefore allowed them to meaningfully assess the options after T_3 and T_4. The results suggest that: (a) students prefer collaborative to individual note-taking, (b) students prefer sharing virtual workspaces in a quiet environment to tangible workspaces in an audible

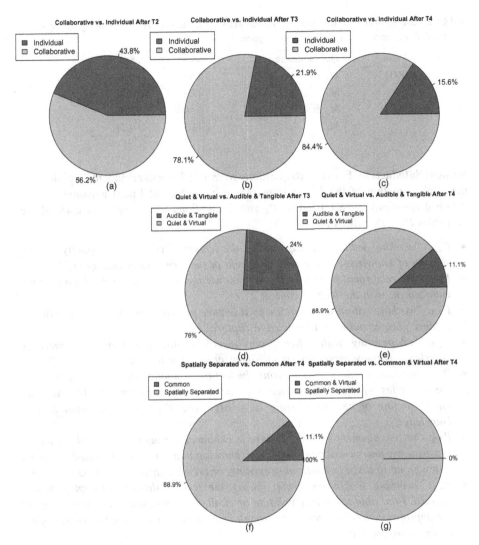

Fig. 7. Distribution of preferred note-taking methods

environment, and (c) students prefer sharing spatially separated workspaces to a constrained common workspace.

Engagement in the Lecture. Table 2 provides the test results of participants' self-assessed engagement from the questionnaires after M_1 (Mean = 2.63, SD = 0.87), after M_2 & M_3 combined (Mean = 3.40, SD = 0.99), and after M_4 (Mean = 4.03, SD = 0.71) using a 5-point Likert scale. T-test results indicated that hypotheses H_1 and M_{21} were confirmed at $p < 0.05$, that is, students are more engaged with collaborative than with individual note-taking, and with sharing spatially separated workspaces than with sharing a common workspace.

Table 2. Engagement in the lecture: collaborative vs. individual and sharing of spatially separated workspaces vs. a common workspace

p-value	Questionnaire	Mean	Std Dev
0.0001	M_1	2.63	0.87
	M_4	4.03	0.71
0.0002	M_2 & M_3 combined	3.40	0.99
	M_4	4.03	0.71

Student Satisfaction. For H_{22}, 16 questions were asked to gather students' opinions on the implementation of the Seven Principles using a 5-point Likert agreement scale, where 4 questions were related to P_2 and 2 questions were related to each of the principles $P_3 - P_7$, which are:

- $P_{2.1}$: *"working within a group during the lecture develops my capacity for the sharing of knowledge to the mutual benefit of the members of that group"*,
- $P_{2.2}$: *"the reciprocal benefit of sharing knowledge within our group increases my engagement with the lecture content"*,
- $P_{2.3}$: *"working within a group during a lecture promotes cooperation in order to achieve greater individual knowledge capture"*,
- $P_{2.4}$: *"cooperating with other group members during a lecture to increase knowledge capture increases my engagement with the lecture"*,
- $P_{3.1}$: *"working within a group during the lecture promotes active learning through the examples provided by myself and other group members to illustrate the understanding of new or existing knowledge and also, the application of that knowledge"*,
- $P_{3.2}$: *"my engagement with the lecture is enhanced through the examples I use to illustrate my new or existing knowledge, knowing that it is to be accessed by others in my group to assist in confirming existing, or creating their own new knowledge"*,
- $P_{4.1}$: *"working within my group during the lecture delivers prompt feedback because I am able to quickly validate or challenge my new/existing knowledge at the appropriate teaching moment through being able to access the knowledge of the other group members"*,
- $P_{4.2}$: *"my engagement during the lecture is enhanced due to the prompt feedback I am able to access from the work of my group members"*,
- $P_{5.1}$: *"working within my group during the lecture emphasises time on task as the shared goal is greater knowledge capture, which requires commitment. The awareness that others will be able to see any reduction in note output such as would occur if I checked out a Facebook page, sent an SMS, fell asleep etc. provides the motivation to remain on task"*,
- $P_{5.2}$: *"my engagement with the lecture is enhanced because I am motivated to remain on task in order that I meet the expectations of my group members"*,
- $P_{6.1}$: *"working within my group encourages me to meet the agreed expectations of the group in terms of meeting our shared goal of increasing the knowledge capture during the lecture"*,

- $P_{6.2}$: *"my engagement with the lecture is enhanced through a joint commitment with my other group members to meet an expected standard",*
- $P_{7.1}$: *"working within a group exposes me to the different talents and ways of learning of others in my group. As I know they have the same commitment to our shared goal my respect for these different methods of knowledge capture is increased",* and
- $P_{7.2}$: *"my engagement with the lecture in enhanced when I am working with others who share the same goal as I and also expose me to different talents and ways of learning which provide a different perspective to my own".*

Figure 8 shows that sharing of spatially separated workspace in M_4 is generally perceived as a better implementation of the Seven Principles than only sharing a common workspace in M_2 or M_3. All T-test results except $P_{2.3}$, $P_{5.1}$, $P_{6.2}$, and $P_{7.1}$ were statistically significant at $p < 0.05$.

For $H_{23} - H_{26}$, 5 questions were asked to gather students' satisfaction with the way collaborative note-taking is supported using a 5-point Likert agreement scale, which are:

- **Helpful:** "access to the work of my group members during the lecture using this method was helpful",
- **Distracting:** "access to the work of my group members during the lecture using this method was distracting",
- **Beneficial.1:** "access to the work of my group members during the lecture using this method provided a greater benefit than working alone",

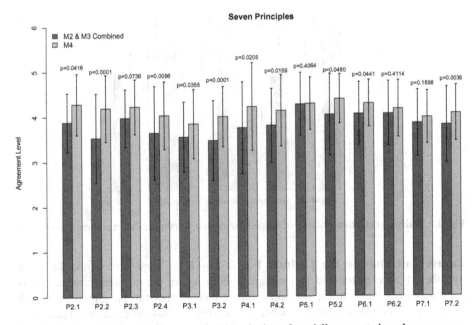

Fig. 8. Implementation of the seven principles: sharing of spatially separated workspaces vs. a common workspace

- **Beneficial.2:** "the benefit from access to the work of my group members during the lecture using this method more than compensated for the level of distraction, if any, it caused me", and
- **Attending:** "if cooperation supported by this method were permitted during non-interactive lectures, I would attend this lecture".

Figure 9 shows that students are more satisfied with sharing spatially separated workspaces in M_4 than only sharing a common workspace in M_2 or M_3. In particular, (a) access to the work of group members through sharing spatially separated workspaces is more helpful than sharing a common workspace ($p = 0.0079$), (b) access to the work of group members through sharing spatially separated workspaces is less distracting than sharing a common workspace ($p = 0.0001$), (c) access to the work of group members through sharing spatially separated workspaces provides a greater benefit than sharing a common workspace ($p = 0.0004$ and $p = 0.0001$), and (d) students are more likely to be persuaded to come to non-interactive lectures with the support of sharing spatially separated workspaces than sharing a common workspace ($p = 0.0013$).

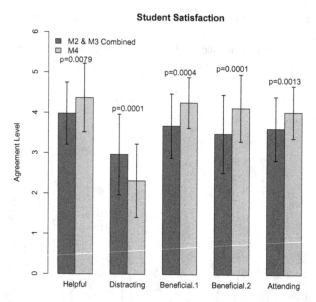

Fig. 9. Student satisfaction: sharing of spatially separated workspaces vs. a common workspace

Engagement Outside the Lecture. For $H_{31} - H_{33}$, 3 questions (one for each hypothesis) were distributed to participants both in the pre-questionnaire and post-questionnaire in order to capture and compare their attitudes to the engagement with topic material outside the lectures, when considering collaborative and individual note-taking methods. Table 3 shows the results, where H_{32} and H_{33} are positively confirmed ($p = 0.0001$). That is, more students would choose to revise notes and discuss notes with at least one other person after the lecture with the support of collaborative note-taking than individual note-taking. However, H_{31}, that is more students would prepare for the lecture beforehand with collaborative than individual note-taking, was not confirmed at the $p < 0.05$ level.

Table 3. Engagement outside the lecture: collaborative vs. individual note-taking

p-value	Questionnaire	Mean	Std Dev
0.06	Pre: H_{31}	2.59	0.95
	Post: H_{31}	3.06	1.01
0.0001	Pre: H_{32}	2.81	1.36
	Post: H_{32}	3.84	0.81
0.0001	Pre: H_{33}	1.88	1.13
	Post: H_{33}	3.16	0.94

The results from the post-questionnaire also revealed the key features of Group-Notes that make social interaction during a lecture viable and that make M_4 distinct from M_3. Of 32 responses,

- 29 (or 90.6 %) quoted the novel multi-user interface supporting spatially separated workspaces,
- 28 (or 87.5 %) quoted the real-time access to other group member's notes, and
- 25 (or 78.1 %) quoted the flexibility of the user interface and the ability to minimise distraction.

The real-time availability of all content generated by group members but accessible in different workspaces according to the needs and abilities of the individual allows for the cooperation and/or collaboration of the group members to provide a benefit over working alone.

6 Conclusions and Future Work

Our approach to increasing student engagement in lectures using mobile devices is innovative, both in terms of the purpose that the device is used for and the techniques allowing for different learning abilities and styles. By providing an interface that permits the individual to view/hide information sources as required, and by allowing them to benefit from and contribute to the shared goals of their self-selected group without negatively impacting their own, we feel that GroupNotes will provide a positive addition to students' individual learning outcomes.

Future work for this project includes capturing whiteboard activity into the work-space using the camera available on most mobile devices. Device agnosticism is another goal and future work will involve the inclusion of other types of devices (the current prototype supports Android tablets). A further area of research is to determine the usability of the designed interface on smartphones, to see what features translate well into the smaller screen surfaces. Future work will also encompass longer term testing to determine the universality of features built into GroupNotes for different types of learners and for different knowledge domains.

We are conscious of the limitations in this work. First, peer interactions may potentially distract students to distraction from the lecturer or the lecture content. We will thoroughly investigate this as well as distraction within a group. Second, self-assessed engagement alone may not accurately reflect their actual level of engagement, therefore we will investigate other more objective means, including those assessing learning outcomes. Third, we will study the impacts of other grouping schemes, including those that combine both strong and weak students or combine self-motivated and under-motivated students. Last, analysis of the quiz results after each test did not show a positive correlation between engagement and learning outcome (that is more engagement is translated to better learning outcome), which is not unexpected given the short testing period (4 lectures in 2–3 weeks). We will conduct longer-term tests to study the extent of the expected increased learning outcomes to students as a result of their increased engagement across all types of lectures.

References

1. Aagard, H., Bowen, K., Olesova, L.: Hotseat: opening the backchannel in large lectures. EDUCAUSE Q. **33**(3), 2 (2010)
2. van Ackere, A.: The principal/agent paradigm: Its relevance to various functional fields. Eur. J. Oper. Res. **70**(1), 83–103 (1993)
3. Bligh, D.A.: What's the Use of Lectures?. Jossey-Bass, San Francisco (2000)
4. Chickering, A.W., Gamson, Z.F.: Seven principles for good practice in undergraduate education. AAHE Bull. **39**(7), 3–7 (1987)
5. Costa, J.C.E., Ojala, T., Korhonen, J.: Mobile lecture interaction: making technology and learning click. In: IADIS International Conference Mobile Learning, pp. 119–124 (2008)
6. Davis, R.C., et al.: Notepals: lightweight note sharing by the group, for the group. In: The SIGCHI Conference on Human Factors in Computing Systems, pp. 338–345 (1999)
7. Falkner, K., Munro, D.S.: Easing the transition: a collaborative learning approach. In: Australasian Computing Education Conference, pp. 65–74 (2009)
8. Hartley, J., Daviesa, I.K.: Note-taking: a critical review. Innov. Educ. Train. Int. **15**(3), 207–224 (1978)
9. Hitchens, M., Lister, R.: A focus group study of student attitudes to lectures. In: The Australasian Conference on Computing Education, pp. 93–100 (2009)
10. Jarvela, S., Naykki, P., Laru, J., Luokkanen, T.: Structuring and regulating collaborative learning in higher education with wireless networks and mobile tools. Educ. Technol. Soc. **10**(4), 71–79 (2007)
11. Kam, M., et al.: Livenotes: a system for cooperative and augmented note-taking in lectures. In: The SIGCHI Conference on Human Factors in Computing Systems, pp. 531–540 (2005)
12. Kiewra, K.A.: Notetaking and review: the research and its implications. Instr. Sci. **6**(3), 233–249 (1987)
13. Koschmann, T.: Toward a dialogic theory of learning: Bakhtin's contribution to understanding learning in settings of collaboration. In: International Conference on Computer Supported Collaborative Learning (1999)
14. Litchfield, A., Raban, R., Dyson, L.E., Leigh, E., Tyler, J.: Using students' devices and a no-to-low cost online tool to support interactive experiential mlearning. In: International Conference on Advanced Learning Technologies, pp. 674–678 (2009)

15. MacArthur, J.R., Jones, L.L.: A review of literature reports of clickers applicable to college chemistry classrooms. Chem. Educ. Res. Pract. **9**, 187–195 (2009)
16. Milrad, M., Jackson, M.H., Bergman, D.: Exploring the potential of mobile services to support learning and communication in university classes. In: Proceedings of IEEE International Workshop on Wireless and Mobile Technologies in Education, pp. 107–112 (2005)
17. Patry, M.: Clickers in large classes: from student perceptions towards an understanding of best practices. Int. J. Scholarsh. Teach. Learn. **3**(2), 17 (2009)
18. Reilly, M., Shen, H.: GroupNotes: encouraging proactive student engagement in lectures through collaborative note-taking on smartphones. In: International Conference on Computer Supported Collaborative Learning, pp. 908–909 (2011)
19. Reilly, M., Shen, H.: Unobtrusive student collaboration during lectures with smartphones. In: Proceedings of the 6th International Workshop on Ubiquitous and Collaborative Computing, pp. 56–65 (2011)
20. Reilly, M., Shen, H., Calder, P., Duh, H.: Understanding the effects of discreet real-time social interaction on student engagement in lectures. In: Proceedings of the 25th Australian Computer-Human Interaction Conference, pp. 193–196 (2013)
21. Reilly, M., Shen, H., Calder, P., Duh, H.: Towards a collaborative classroom through shared workspaces on mobile devices. In: Proceedings of the 28th British Human Computer Interaction Conference, pp. 335-340 (2014)
22. Savill-Smith, C.: The use of palmtop computers for learning: a review of the literature. Brit. J. Educ. Technol. **36**(3), 567–568 (2005)
23. Shen, H., Reilly, M.: Personalized multi-user view and content synchronization and retrieval in real-time mobile social software applications. J. Comput. Syst. Sci. **78**(4), 1185–1203 (2012)
24. Simon, B., Davis, K., Griswold, W.G., Kelly, M., Malani, R.: Noteblogging: taking note taking public. In: The SIGCSE Technical Symposium on Computer Science Education, pp. 417–421 (2008)
25. Simon, H.A.: Theories of bounded rationality. In: McGuire, C.B., Radner, R. (eds.) Decision and Organization, pp. 161–176. North-Holland Publishing Company, Amsterdam (1972)
26. Steimle, J., Brdiczka, O., Muhlhauser, M.: Coscribe: using paper for collaborative annotations in lectures. In: International Conference on Advanced Learning Technologies, pp. 306–310 (2008)
27. Tinzmann, M., Jones, B., Fennimore, T., Bakker, J., Fine, C., Pierce, J.: What is the collaborative classroom? Technical report, North Central Regional Educational Library (1990)
28. Ward, N., Tatsukawa, H.: A tool for taking class notes. In. J. Hum.-Comput. Stud. **59**(6), 959–981 (2003)

Physical Spatial Interaction

Cataloguing Physicality Values Using Physical Quantitative Evaluation Method

Mahmood Ashraf[1] and Masitah Ghazali[2(⊠)]

[1] Department of Computer Science, Federal Urdu University of Arts,
Science and Technology, Islamabad, Pakistan
mahmood@fuuastisb.edu.pk
[2] Software Engineering Department, Universiti Teknologi Malaysia,
Skudai, Malaysia
masitah@utm.my

Abstract. Physical interfaces suffer from interaction complexities leading to usage difficulty and poor acceptance by the end-users. Usability techniques focus on the overall usability issues while overlooking the in-depth physicality aspects of the interface and interaction. This study proposes a physicality-focused quantitative evaluation method (PQEM) to assist embedded system developers in managing the interaction complexities of their products. The acceptance of the embedded system developers towards the proposed method was assessed by means of a user study. The results suggested their strong acceptance. By using this method, we further propose a range of values for appliances, which can be treated as a catalogue or as guidelines as they design and develop physical interfaces. A second user study was conducted to assess the acceptance of embedded system developers for the proposed catalogue. In this study, spatial cognition is concerned in assisting and facilitating the designers and developers in designing physical interfaces of embedded products. We used intensive interviewing as the cognitive method to assess the influence of the catalogue of values from PQEM on performance in designing and developing physical interfaces.

Keywords: Physicality · Embedded systems · Evaluation · Intuitive interaction · Spatial cognition

1 Introduction

Contemporary works have verified that the traditional concept of usability might be insufficient especially concerning product quality and user satisfaction [1–3]. In the multimodal quality taxonomy proposed by Wechsung [1], he placed emphasis on spatial cognition and incorporated it at various levels of the taxonomy. This underlying study supports the proposed taxonomy of Wechsung [1] in terms of relationship between spatial cognition and physicality. This study advocates that the physicality aspects affect the spatial cognition. Wechsung also divided user abilities into spatial cognitive abilities and physical capabilities as these work together and influence each other [1]. This study expands on the physicality aspects of interfaces. When the interleaved sequence of both mental operations and physical actions are performed, these produce more efficient interaction than if one of these are used individually [4].

© Springer International Publishing Switzerland 2015
T. Wyeld et al. (Eds.): OzCHI 2013, LNCS 8433, pp. 197–214, 2015.
DOI: 10.1007/978-3-319-16940-8_10

The physical user interfaces of embedded systems are a major part of the physical environment [4–6]. It is repeatedly reported that the physical environment heavily affects the cognitive load and learning of users [7–9].

It has been empirically proven that users try to apply various cognitive mechanisms for interacting with space [10]. The embodied interfaces provide a strong platform for engaging spatial ability [11]. However, to date, this area has not been sufficiently addressed [11]. There is a need for design guidelines for designers of embodied tangible products especially covering the spatial cognitive ability [11]. Efficient spatial cognitive strategies can be achieved by exploiting the benefits of physical interfaces including 3D manipulation space, eyes-free tactile feedback and offline workspace [4].

In this study, spatial cognition is concerned in assisting and facilitating designers and developers in designing the physical interfaces of embedded products. In doing so, a catalogue is presented to them, which consists of values of embedded products. We use an intensive interview as the cognitive method to see as to whether the catalogue, in some ways help them performing better in designing and developing physical interfaces.

The next section provides an overview of the physicality concept. The following section outlines the physicality quantitative evaluation method (PQEM) including an example. A user study is first presented to evaluate the PQEM. Subsequently, the concept of cataloguing the physicality values based on the PQEM is explained. The second user study is described and its preliminary results are presented. Finally, we discuss and conclude the overall the work.

2 Physicality

Physicality advocates the significance of physical interaction aspects and provides solutions related to embodiment [12–15]. It deals with the engagement and embodiment of the human body and physical artefacts during interaction [16]. According to Norman [15], physicality is the return to physical and mechanical controls that are now equipped with embedded and intelligent processors and also offer communication facilities. Recently, the significance of physicality principles in user interaction has been further recognized [17–20].

Therefore, interacting with embedded systems – especially daily use appliances – requires the consideration of physicality aspects in addition to usability and other aspects [12, 16]. Some of the important physicality aspects or principles include exposed state, controlled state, tangible transitions, bounce-back, inverse action and compliant interaction [6, 13, 16]. Currently, these aspects can only be investigated qualitatively by experts in the field based on their expertise and knowledge [12]. This limitation or dependency slows down the large-scale acceptance and implementation of these principles. In this study, we present a method for quantitatively evaluating the physicality principles so that non-experts of HCI in the software development field can also benefit by introducing these principles into their products. This will help the large-scale applicability of physicality aspects and facilitate the introduction of natural interaction in embedded systems.

Ghazali [6] and Dix et al. [16] investigated mundane physical artefacts and the causes of naturalness and fluidity in interaction. These studies were based on analyses of the physical-digital mappings and design features of mundane home appliances and their controls. These mappings, design features, and the implicit design characteristics were found to significantly affect natural interaction. The cognitive investigation of these characteristics and physical-digital mappings resulted in the discovery of physicality principles. Six important physicality principles are included in the method proposed in this study as mentioned above: exposed state, controlled state, tangible transition, bounce-back, inverse action, and compliant interaction. This paper expands on the physicality principles discovered by Ghazali [6].

Exposed state is the direct and visible mapping between the device's physical and logical states. For example, the shape of light switch in 'on' or 'off' state reveals the logical state. Controlled state is the limitation imposed by the device disallowing the user to return the physical state of the control to its original position. For example, the insertion handle of a conventional toaster locks itself down in the 'on' state. Tangible transition offers tactile feedback to the user on each state transition. It is often observed in dials and other controls, especially rotary controls. In the bounce-back effect, the physical state of the control returns to its original position over time, irrespective of the logical state. For example on/off buttons normally have this effect. Inverse action is the inverse logical effect assigned to physically opposite states - for example, the increase and decrease of values in respectively clockwise and anti-clockwise rotation of various controls. Compliant interaction is the symmetrical aspect of the user and system interaction. For example, the washing machine program dial takes control from the user until the end of washing operation.

With the advent of technology, the mundane objects around us do not remain mundane any more. Computation, decision-making, multi-functionalities, and intelligence are being increasingly embedded in physical artefacts. These daily use devices cannot afford heavy manuals or twenty-four hour online help [15]. Therefore, it is important to apply and benefit from the yet undiscovered benefits of physicality.

3 Physicality Quantitative Evaluation Method (PQEM)

This section introduces the proposed physicality quantitative evaluation method. Presently, the physical user interfaces of embedded systems can be properly evaluated only by experts of HCI and especially experts in the physicality field. However, due to the gap between the software engineering and HCI fields, embedded software developers have been unable to collaborate with HCI experts [21–32]. This practice results in either inadequate evaluation of products by non-experts or unrelated personnel, or release of products without any evaluation of the physical user interface. Therefore, we aim to develop a quantitative evaluation method for embedded system developers according to physicality aspects.

Each control can be assigned three numeric values for a particular principle, that is, zero (0), one (1), or two (2) listed in Table 1.

This range is based on the occurrence level of a physicality principle in that particular control. A value of zero indicates that the particular principle is absent in a given control, while a maximum value of two indicates that the particular principle is fully present in that control. A value of one indicates that the control offers partial

Table 1. Values of physicality principles according to their existence in a control.

Principle	Value	Description
Exposed state (ES)	0	Absent
	1	Partially supported i.e., half or few states exposed
	2	Fully supported i.e., all exposed states
Controlled state (CS)	0	Absent
	1	Partially supported i.e., presence of intermediate state
	2	Fully supported i.e., no intermediate state
Tangible transitions (TT)	0	Absent
	1	Partially supported i.e., unequal number of tangible feedbacks offered than the visible scale or the underlying logical states
	2	Fully supported i.e., all tangible feedbacks equally mapped to the visible scale or the underlying logical states
Bounce-back (BB)	0	Absent
	1	Weakly supported i.e., delay and play on the return path (soft bounce-back)
	2	Strongly supported i.e., no delay at return (hard bounce-back)
Inverse action (IA)	0	Absent
	1	Vague and/or unequal number of mappings at each side
	2	Clear and/or equal number of mappings at each side
Compliant interaction (CI)	0	Absent
	1	Weak compliance i.e., no means available for user to go back during symmetric motion
	2	Strong compliance i.e., user has some means available to go back

features of a particular principle. This means that if the user is only allowed to perform one of the two actions then we say that the exposed state property half exists. For example, if a control offers a toggle function with two state transitions but the user can only afford one of these transitions, then the control exhibits a half exposed state.

The quantitative evaluation of physicality aspects in a particular physical user interface can be performed by using the values assigned in Table 1. It is also significant for the evaluator to check whether a principle is indeed applicable for a particular control according to the given requirements. In cases where a principle is not applicable, the evaluator will assign a mark of 'NA' (stands for Not Applicable). For example, a compliant interaction principle is not possible in a toggle switch button or a simple push button. A template of the evaluation chart with dummy values is shown in Table 2.

An interface may have a number of controls of various types. Each control is evaluated for six physicality aspects. After assigning the values, the Sub-total is calculated (ignoring all the 'NAs'). The Sub-total is the sum of values that represent the existence of a single principle in all controls of the interface. The value of Total for the existence of all physicality principles in an interface is calculated by adding the

Table 2. Template of evaluation chart for any interface (*dummy values).

			ES	CS	TT	BB	IA	CI
I	Control	Type	j = 6					
1	Control1	Lever	2	2	0	2	2	0
2	Control2	Dial	2	0	2	0	2	2
3	Control3	Button	1	0	2	2	0	NA
...	Control...	Switch	1	0	2	2	0	NA
N	Control n	Handle	0	0	0	1	0	0
	Sub-Total		7	2	6	7	4	2
	Total	28*						

sub-totals of all principles. The value of Total can be calculated by applying the following formula (1) on Table 2:

$$Total = \sum_{(i,j)=(1,1)}^{(n,p)} x(i,j) \qquad (1)$$

Where i is the counter for controls, j is the counter for physicality principles, n is the maximum number of controls on an interface, and p is the total number of physicality principles addressed.

Subsequently, the Physicality Score of an interface can be calculated by taking the average based on the number of controls present in that interface:

$$Physicality\ Score = \frac{Total}{(n*p) - N} \qquad (2)$$

where N is the number of 'Not applicable (NA)' instances in the evaluation chart. This Physicality Score, in addition to providing the value of physicality in an interface, can also be used for various other purposes including comparison with standard values or comparison between different interfaces. To better understand the proposed method and the application of physicality principles let us consider the example of the household toaster (Fig. 1) to which we can assign values for the physicality principles from Table 1.

Fig. 1. Physical user interface of a toaster.

The toaster has a total of five controls consisting of: a lever for inserting/ejecting toasts and high-lift feature; a dial for browning settings; and four buttons for miscellaneous functions.

On referring to Fig. 1, the exposed state is the visible and direct mapping between the physical-logical states is possible. The insertion/ejection lever exposes the press-down feature but does not expose the press-up movement to the user - that is, the high-lift feature. Therefore, it offers a partially exposed state and hence can be rated with value one. The browning setting dial has no exposed state features mainly due to the absence of any kind of scale. The current value indicator (in circle) does not expose (indicate) whether the value is high or low, minimum or maximum, or in which rotational direction it will increase or decrease. Hence, a zero value is assigned. The rest of the four button controls are merely surface buttons. These neither clearly expose their push-in feature (in both pressed-in and pressed-out states) as well as any indication about the current logical state from their physical state, hence a value of one is assigned to all. The designers of the toaster have tried to cover this by introducing LEDs.

Controlled state is the limitation imposed by the device preventing the user from returning the physical state of the control to its original position. The main lever of the toaster offers this property (for inserting the toast in the toaster) when the lever is pressed down. Here the lever stays at the final pressed position until the toasting finishes. At finishing time, the lever pops automatically up to the default position (with the toast). Therefore, a value of two is assigned. Additionally, the user can cancel the function any time by pressing the cancel button. The dial is a freely moving control without any restriction imposed for the user. Likewise, the rest of the controls do not offer this principle and hence are rated zero.

Table 3. Physicality values of toaster interface.

i	Control	Type	ES	CS	TT	BB	IA	CI
1	Insertion/Ejection	Lever	1	2	2	2	2	0
2	Browning setting	Dial	0	0	0	0	2	0
3	Bagel	Button	1	0	1	1	0	NA
4	Defrost	Button	1	0	1	1	0	NA
5	Reheat	Button	1	0	1	1	0	NA
6	Cancel	Button	1	0	1	1	0	NA
	Sub-total		5	2	6	6	4	0
	Total	23						

Tangible transition provides the user with tactile or physical feedback on each state transition. The insertion/ejection lever of the toaster offers this principle when it is fully pressed down and locked by the system. Additionally, along with the down movement, the increased tactile resistance also augments the effect; hence, it is rated with value two. The browning setting dial does not offer any tangible transition hence rated zero. The buttons offer a partial tangible transition at the fully pressed-in state hence it is rated one.

In the bounce-back effect, the control returns to its original physical state irrespective of the logical state. The insertion/ejection lever offers bounce-back if released anywhere between the starting and fully pressed positions. The browning setting dial does not offer this principle, while the buttons offer partial bounce-back; that is, they are not clearly visible, and therefore are rated as zero and one, respectively.

The inverse action is the inverse logical effects assigned to opposite physical states. Both the insertion/ejection lever and the browning settings dial offer inverse actions hence rated two, while none of the other controls offer inverse action hence rated zero.

Compliant interaction is the symmetrical or mutual control of a user and a system. However, in the subject a toaster, none of the controls offer this principle. It should be noted that it is possible for the lever and the dial to offer compliant interaction principle for certain use but it is not possible for the existing buttons to offer compliant interaction. Therefore, for lever and dial a value of zero is assigned while NA is assigned for the four buttons.

All values for the toaster are listed in Table 3. Formula (1) is applied to the recorded values as follows (while leaving the NAs):

$$Total = (1 + 0 + 1 + 1 + 1 + 1) + (2 + 0 + 0 + 0 + 0 + 0)$$
$$+ (2 + 0 + 1 + 1 + 1 + 1) + (2 + 0 + 1 + 1 + 1 + 1)$$
$$+ (2 + 2 + 0 + 0 + 0 + 0) + (0 + 0)$$

$$Total = 5 + 2 + 6 + 6 + 4 + 0 = 23$$

We get the value of Total as 23 and by using the Formula (2):

$$Physicality\ Score = \frac{Total}{(n*p) - N} = \frac{23}{(6*6) - 4} = 0.178$$

Hence, the Physicality Score of 0.178 is calculated for the subject toaster interface. These values and score may help embedded system developers to check the physicality aspects of their products.

4   User Study I

This section reports on a user study with the aim of checking the acceptance level of developers for the proposed physicality quantitative evaluation method. The study followed the approach of [6, 33]. In this study, three embedded systems are used; fan; sound system; and oven. These devices have physical user interfaces consisting of a variety of controls (close-ups in Fig. 2). The fan has five controls; a dial and four buttons. The sound system has four controls; a push button and three dials. The oven has three controls; a toggle button and two dials.

Method. The participants first completed a pre-test questionnaire consisting of their background and preferences. Then they were briefed about the physicality principles with examples. The subject devices were then introduced to them and they were asked

Fig. 2. Close-ups of a fan, sound system and oven interfaces, respectively.

to become familiar with them. Subsequently, the participants were briefed about the proposed quantitative evaluation method. Afterwards they conducted the evaluation according to the proposed method and recorded their ratings on the evaluation forms. After the evaluation, their acceptance level was recorded by means of a post-test questionnaire. The participants were presented with three interfaces one by one in random order to avoid sequential effects.

Participants. The participants were solicited voluntarily without any benefits from an embedded software engineering group. The participants had a variety of software and embedded system development experience ranging from two to sixteen years. A total of 23 participants were recruited, consisting of 15 males and 8 females having diverse cultural, ethnic and national backgrounds.

Measures. Both qualitative and quantitative data were collected. Written notes were also taken on participants' behaviour and other issues by the relevant investigator. The participants were not time-bound and could work at their own pace in the most comfortable way.

Procedure. The study took five days in one of our department's research labs. A single investigator was responsible for briefing and managing questionnaires. The study was conducted in four stages.

In the first stage, the participants completed a pre-test questionnaire with information about their personal and professional experiences. The questionnaire also included questions about their expected preferences about any evaluation method in general (as they were not introduced to the PQEM at this stage of the study).

In the second stage, the participants were briefed about the physicality principles with examples of embedded system interfaces followed by a question and answer session to clarify any ambiguities. Later, the participants were presented with the three subject devices (fan, sound system and oven). The participants were asked to simply play with the devices to gain familiarity (Fig. 3). Subsequently, they were briefed about the proposed quantitative evaluation method followed by a question and answer session to resolve any confusion surrounding the evaluation method.

In the third stage, an evaluation form was given to them for recording the physicality values of subject interfaces and the participants undertook the evaluation process. At the end of the evaluation process, they were presented with a post-test questionnaire to report their feedback for the proposed method. During the study, the investigator was

Fig. 3. Participants interacting with devices during case study.

available to answer any queries. Finally, the collected data was analysed by the investigator.

Data Collection. All the questionnaires and forms were collected on an individual basis directly by the investigator. The indirect data were recorded through observation notes. Additionally, the questionnaires contained a comments section that was filled in by the participants for the qualitative analysis. The observation notes and especially the comments helped the investigators in qualitative analysis. The recorded data revealed the performance patterns of participants.

Results. There were various categories of results from the study. In this paper, the focus is on the acceptance level of participants towards the proposed method. According to the information recorded in the pre-test questionnaire forms, ten (10) participants already knew about physicality and had tried to incorporate it in their products. Three of the participants had experience of conducting other evaluations of physical user interfaces.

The majority of participants (17 of 23) expected three properties in any evaluation method in general i.e., (i) easy to conduct, (ii) consume a little time to perform, and (iii) can have the ability to produce correct results. Two participants preferred the evaluation method to have the first two properties and the remaining three participants preferred one of each property, respectively. The majority of the participants (22 of 23) preferred to have quantitative results as an evaluation method for physical user interfaces of embedded systems.

Participants were informed before the experiment that they were participating in a study that would assess a quantitative evaluation method for checking the physicality aspects in embedded system interfaces. A post-test questionnaire was used for the measurement of the degree to which the method was acceptable. The acceptance level results are divided in two categories: quantitative and qualitative.

Quantitative Results. Each participant applied the proposed method on all three embedded systems and was presented with the acceptance level questionnaire at the end of each evaluation. Therefore, 23 participants performed an evaluation of the proposed method for a total of 69 times. The questions in the acceptance level questionnaire and the results are shown in Fig. 4 and are based on the 69 results on a scale from zero (strongly disagree) to 5 (strongly agree).

As per the expectations of the participants provided in the pre-test questionnaires, the participants found the method easy (question 1), consumed little time to perform

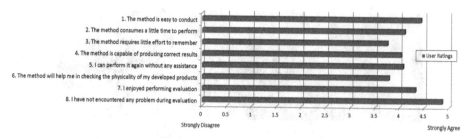

Fig. 4. Result of acceptance level of proposed method by participants (Bars represent the mean of the numerical scores for each question).

(question 2), and capable of producing correct results (question 4). Participants also reported the method could be performed without assistance (question 5). Participants hardly faced any problems during the study as evident from the highest ratings for question 8. Interestingly, the top three rated parameters also included the enjoyment factor (question 7). This shows that the participants were keen to use the method. The participants agreed that the method required little effort to remember and was applicable to the development of products, but these were rated less compared to others. Question 3, 'the method requires little effort to remember', was rated at 3.75 showed that the participants did not find it effortless to remember the method. This was due to the issue of learnability of six principles, especially in the first attempt. Question 6, 'This method will help me to check the physicality aspects of my developed products', was also rated at 3.75 showed that the participants were not very confident whether they would be able to apply the method in their products at their work place. According to the participants' comments, there existed administrative and managerial issues at their work places that may hinder the application of PQEM. These reasons have been previously discussed in Hirasawa et al. [47]. They are also referred to in the next section.

Additionally, we analysed the gender-based differences but it was revealed that the acceptance level was approximately equal for both groups. Similarly, the sequence of devices presented to the participants showed a subtle effect on the results. In cases when the fan or oven was presented first, the acceptance level results were slightly better than the cases when the sound system was presented first. However, this difference was negligible. Overall, the quantitative results show that the participants appreciated the proposed method.

Qualitative Results. The participants recorded their descriptive comments in the questionnaire forms. This provided us valuable feedback and helped us to understand the reasons for the quantitative results. One of the issues was that the participants did not know about physicality principles before the study. They were briefed about the principles in the initial stages of study. However, they encountered some problems in distinguishing the principles from each other. They explicitly reported that the method itself was very easy and did not require them to remember much but the results reflected in the quantitative analysis (question 3) were due to the initial effect of the memory requirement for remembering the principles. As one of the participants commented, *"I always have forgotten the key methods [principles]"*. However, with practice, this parameter improved as the results of later evaluations had better results than the first evaluation. For instance, one of the participants after her third evaluation commented, *"...now getting easier"*.

Another issue was in relation to question 6, which was about the potential application of this method in their products. The quantitative results were possibly affected (comparatively low) because they depended on a number of constraints, including the specific nature of their products, their company's flexibility in incorporating new methods, and pragmatic time limitations. These business and management-related problems are major obstacles when applying new tools and techniques in small or especially medium and large organizations. Such issues have been reported in various other studies, especially in the study of Hirasawa et al. [47], but were out of the scope of this research. One of the participants commented, "In my opinion, the method is easy and I believe this method is capable to produce [the] correct result". Another participant commented, "Nice, easy, simple, quick way of evaluating". At the end of the study, many participants asked the investigator to call them again for another study. Three of the participants requested a second opportunity because they had enjoyed it.

5 Cataloguing the Physicality Values

To systematically organize the physicality values of interfaces, we created a catalogue. The catalogue concept was chosen because it could help in relieving the cognitive burden of designers and developers [34]. By featuring ranges of values of physical interfaces a catalogue can assist embedded developers to understand and quickly recognize what makes a product usable or physically-intuitive (physintuitive). As mentioned earlier in this paper, physicality and spatial cognition aspects are closely associated and work together. When the interleaved sequence of both mental operations and physical actions are performed, these produce more efficient interactions than if one of these are used individually [4].

In this section, we will describe how we derived the catalogue by using the PQEM, which was described earlier in the previous section. Three HCI researchers analyzed seven types of home appliances based on the PQEM. These seven types include oven, juicer, toaster, rice cooker, table fan, kettle, and egg beater. A number of interfaces on an appliance type were analyzed. Only the primitive functions of the appliances are selected.

The analysis was then translated into the form of a catalogue of recommendations for the assistance of appliance designers. The designers can check whether the controls they are using meet the recommendations in the catalogue. The catalogue mainly recommends two aspects, including the type of controls and their physicality values. Table 4 displays the catalogue.

Table 4 also lists the appliance types, selected primitive functions, the recommended controls for primitive functions, recommended physicality values for each physicality principle, control-wise total value and appliance-wise total value. For example, the first appliance type i.e., oven is recommended to have two dials for different functions (temperature and timer settings). However, it is important to note that their recommended physicality values are different due to the dissimilar nature of their functions. The timer control must have controlled state and compliant interaction properties.

Table 4. Catalogue for the designers.

Appliance type	Selected primitive function	Recommended control	Recommended physicality values						Total (control wise)	Total (appliance wise)
			ES	CS	TT	BB	IA	CI		
Oven	Temperature	Dial	2	NA	2	NA	2	NA	6	16
	Timer	Dial	2	2	2	NA	2	2	10	
Juicer	Start/Stop	Toggle button	2	NA	2	NA	2	NA	6	12
	Pulse	Pulse	2	NA	NA	2	2	NA	6	
Toaster	Insertion/ Ejection	Lever	2	2	2	NA	2	NA	8	22
	Browning settings	Dial	2	NA	2	NA	2	NA	6	
	Cancel/Stop	Push button	2	NA	2	2	2	NA	8	
Rice cooker	Start/Stop	Toggle button	2	NA	2	NA	2	NA	6	12
	Function selector	Toggle button / Dial	2	NA	2	NA	2	NA	6	
Table fan	Start/Stop	Toggle button	2	NA	2	NA	2	NA	6	12
	Speed selector	Dial	2	NA	2	NA	2	NA	6	
Kettle	Start/Stop	Toggle button	2	NA	2	NA	2	NA	6	6
Egg beater	Start/Stop	Toggle button	2	NA	2	NA	2	NA	6	12
	Speed selector	Dial	2	NA	2	NA	2	NA	6	

The strong physicality of dial controls is shown by a high total control-wise value i.e., 10. In another example, it is recommended that the rice cooker can either offer a toggle button in case of two settings or a dial in case of 3 or more settings. Both controls should offer the same physicality values.

6 User Study II

This section reports a second user study with the aim of checking the acceptance level of embedded developers for the proposed physicality catalogue that was derived from using the PQEM.

Method. Unlike the first user study, the second user study used an intensive interview method. The participants were first reminded about the PQEM from the first user study. Subsequently, the participants were briefed about and presented with the catalogue of physicality values and its usage. The participants were asked to carry out a short exercise, i.e. to design an embedded device, using the catalogue. Their acceptance level was recorded from the interview.

Participants. The participants from the first user study were maintained. As the participants already know the working of the PQEM, it was easier for them to report their feedback for the current application of PQEM i.e., cataloguing the physicality values.

Measures. We used only quantitative procedures in our interview. The participants were not time-bound and could work at their own pace in the most comfortable way.

Procedure. The study took place over a period of three weeks. A single investigator was responsible for contacting participants, briefing, and managing questionnaires. The study was conducted in three stages, as participants' particulars are the same from user study I. In the first stage, the same participants from user study I were contacted and briefly informed about the extension work on PQEM to get their approval in participating in the user study II.

In the second stage, the participants were reminded about the PQEM, and briefly informed about the catalogue of physicality values and its usage. The participants were asked to look at the catalogue and to apply it in the design, informally: i.e. by doing a simple design exercise .In the third stage, the participants provided their feedback by filling the questionnaire.

Data Collection. The data were collected from the study, followed by an analysis.

Results. Overall, participants showed their confidence in the catalogue. The results (averaged values) are shown in Fig. 5.

Firstly, participants mostly agreed (4 out of 5) that the catalogue is helpful in enhancing the physicality aspects of the products. Secondly, they agreed (4.25 out of 5) that the catalogue makes it easy to check the physicality aspects. The highest rating (4.5 out of 5) is for their confidence in using the catalogue to design products. Participants also agreed (4 out of 5) that the recommended physicality values reduce the design time.

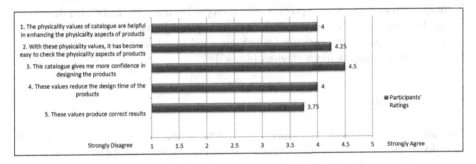

Fig. 5. Participants' ratings for the catalogue.

Finally, participants were less convinced that the catalogue values produce correct results (3.75 out of 5). Nonetheless, this value is well above the middle rating of "agreed" i.e., value 3. Therefore, overall the participants showed their confidence in the catalogue.

7 Related Work

While some studies have investigated the physical user interfaces and their interactions relating to physicality, we have not found any that addressed the in-depth physicality aspects of physical interfaces of embedded systems. The closest to our work is a study by Meier et al. [35] that investigated the way people used thermostats at homes and discovered various interaction complexities. However, the reasons behind such problems that they identified were mainly affordance and usability issues while failing to thoroughly address physicality aspects. Another study by Carvalho and Soares [36] analyzed the ergonomic and usability aspects of automobile dashboards but their work was also limited to the problems regarding visual interface (that is: layout management; lightening; colors; visual signs and icons), while again failing to address the physicality of controls.

Related studies include a series of works by Blackler et al. [37, 38], Lawry et al. [39], and Reddy et al. [40], who empirically evaluated the major effects of cognitive abilities and technology familiarity on the intuitiveness of interaction with electronic devices. However, they focused on the factors of user age, prior experience and the appearance of the products while not addressing the physicality. Likewise, Biswas and Langdon [41], Langdon et al. [42], Persad et al. [43], and Wilkinson et al. [44] provided designers with an approach which they can employ in the design process to match their understandings and responses to product features with that of the users'. Their proposed modeling approach compares the similarities and differences between user and designer models, checks the degree of compatibility, and makes certain design decisions. Although the aim of those works relates to ours, its targeted domain and solution are not concerned directly with physicality but rather the mental models of users and designers, aging and prior experience.

There are various ways to assist the designers and developers in their activities as to facilitate them in the cognitive process. We attempted to assist the embedded designers further by offering them the catalogue. The concept is similar to the ones with categorizing information systematically, which are used as guidelines or as checklists to make the design process more efficient [45, 46].

8   Conclusion

The importance of physicality aspects is evident in the physical user interfaces of embedded systems. Physicality aspects affect the spatial cognition of users [1]. Embedded system designers are trying to design user-friendly interfaces but usability methods are insufficient due to the physicality aspects of interfaces. Embedded system designers need to address the physicality aspects of interfaces in order to improve the spatial cognition. Following the principles of physicality serves to produce intuitive interaction. Each of these physicality principles (exposed state, controlled state, tangible transition, bounce back, inverse action, and compliant interaction) improves spatial cognition. This relationship is also suggested by Wechsung [1]. The controls of the device and the human body should speak the same language. These physicality aspects recruit our inherent and intrinsic cognitive and physical capabilities. Therefore, it is necessary for the embedded system developers to incorporate physicality aspects in their products. In this work, we have discussed the proposed method in detail using an example of embedded system device (a toaster). A user study was conducted for assessing the degree to which the method was accepted by developers. The results of the study showed that the embedded system developers found the proposed method easy, quick, effective and reliable. The participants of the study showed good interest in the method. Moreover, this work also discovered that the principles of exposed state and inverse action were found to be more significant in producing intuitiveness. The proposed method can be applied to a number of embedded systems with physical interfaces. Therefore, we recommend and encourage embedded system developers to introduce physicality principles in their products for intuitive interaction.

In line with Wechsung [1], this study suggests that the spatial cognition is concerned in assisting and facilitating designers and developers in designing physical interfaces of embedded products. This study has extended the proposed physicality quantitative evaluation method (PQEM) as a catalogue of physicality aspects. The proposed catalogue, which facilitates the spatial cognition, consists of mainly two recommendations physintuitive controls and physicality values for appliances.

In order to check the level of designers' acceptance of the proposed catalogue, we used intensive interview to see whether the categorization of values (from PQEM) helps designers perform better in designing and developing physical interfaces. The preliminary results are promising and the time designers had to spend in performing the PQEM was further reduced. Building on the concept of Wechsung [1] about physicality and spatial cognition, this work has proposed a physicality-focused quantitative evaluation method (PQEM) and a catalogue of physicality to assist embedded system designers in managing the interaction complexities of their products. This catalogue, which supports the spatial cognition, is a kind of guideline for the designers of physical interfaces. On evaluation by embedded system designers, the proposed methods were well received. From the results, two aspects need to be addressed further - namely: the learning requirement for the six principles at the first use of PQEM and the management issues in practicing the PQEM in organizations. These issues will be addressed in future studies Therefore, the objective of this series of work is to motivate designers that their designed physical interfaces should be physintuitive and not just intuitive.

Acknowledgements. This work is supported by Universiti Teknologi Malaysia, Skudai, Malaysia.

References

1. Wechsung, I.: What to evaluate? a taxonomy of quality aspects of multimodal interfaces. In: Mukhopadhyay, S.C. (ed.) An Evaluation Framework for Multimodal Interaction T-Labs Series in Telecommunication Services 2014, pp. 23–46. Springer International Publishing, Switzerland (2014)
2. Norman, D.: The way I see it: looking back, looking forward. Interactions **17**(6), 61–63 (2010)
3. Kim, J., Han, S.H.: A methodology for developing a usability index of consumer electronic products. Int. J. Ind. Ergon. **38**(3), 333–345 (2008)
4. Antle, A.N., Wang, S.: Comparing motor-cognitive strategies for spatial problem solving with tangible and multi-touch interfaces. In: Proceedings of the 7th International Conference on Tangible, Embedded and Embodied Interaction(TEI 2013), pp. 65–72. ACM, New York (2013)
5. Antle, A.N., Corness, G., Droumeva, M.: What the body knows: exploring the benefits of embodied metaphors in hybrid physical digital environments. Interact. Comput. **21**(1–2), 66–75 (2009)

6. Ghazali, M.: Discovering Natural Interaction of Physical Qualities to Design Fluid Interaction for Novel Devices. Research Monograph, Universiti Teknologi Malaysia, Malaysia (2007). ISBN 978-967-353-691-7

7. Choi, H., van Merriënboer, J.J.G., Paas, F.: Effects of the physical environment on cognitive load and learning: towards a new model of cognitive load. Edu. Psychol. Rev. 26(1), 225–244 (2014)

8. Nathan, M.J.: An embodied cognition perspective on symbols, gesture and grounding instruction and embodiment. Debates on Meaning and Cognition, Vol. 18, pp. 375–396. Oxford, UK (2008)

9. Wilson, M.: Six views of embodied cognition. Psychon. Bull. Rev. 9(4), 625–636 (2002)

10. Lozano, S.C., Hard, B.M., Tversky, B.: Putting action in perspective. Cognition 103(3), 480–490 (2007)

11. Clifton, P.: Designing embodied interfaces to support spatial ability. In: Proceedings of the 8th International Conference on Tangible, Embedded and Embodied Interaction (TEI 2014), pp. 309–312. ACM, New York (2014)

12. Dix, A., Ghazali, M., Ramduny-Ellis, D.: Modelling devices for natural interaction. Electron. Notes Theoret. Comput. Sci. 208, 23–40 (2008)

13. Ghazali, M., Dix, A.: The role of inverse actions in everyday physical interactions. In: Proceedings of the 2nd International Workshop for Physicality, pp. 21–26 (2007)

14. Hare, Joanna., Gill, S., Loudon, G., Ramduny-Ellis, D., Dix, A.: Physical fidelity: exploring the importance of physicality on physical-digital conceptual prototyping. In: Gross, T., Gulliksen, J., Kotzé, P., Oestreicher, L., Palanque, P., Prates, R.O., Winckler, M. (eds.) INTERACT 2009. LNCS, vol. 5726, pp. 217–230. Springer, Heidelberg (2009)

15. Norman, D.: The next UI breakthrough, part 2: Physicality. Interaction vol. 14, pp. 46–47. ACM, New York (2007)

16. Dix, A., Ghazali, M., Gill, S., Hare, J., Ramduny-Ellis, D.: Physigrams: modelling devices for natural interaction. Formal Aspect Comput. 21, 613–641 (2009)

17. Ashraf, Mahmood, Ghazali, Masitah: Towards Natural Interaction with Wheelchair Using Nintendo Wiimote Controller. In: Zain, Jasni Mohamad, Wan Mohd, Wan Maseri bt, El-Qawasmeh, Eyas (eds.) ICSECS 2011, Part III. CCIS, vol. 181, pp. 231–245. Springer, Heidelberg (2011)

18. Ashraf, M., Ghazali, M.: Augmenting intuitiveness with wheelchair interface using Nintendo wiimote. Int. J. New Comput. Archit. Appl. 1, 1000–1013 (2011)

19. Ashraf, M., Ghazali, M.: Interface design for wheelchair using Nintendo Wiimote controller. In: Proceedings of the 2nd International Conference On User Science And Engineering: Beyond Usability (i-USEr2011), pp. 54–59. IEEE. Malaysia (2011)

20. Ashraf, M., Ghazali, M.: Investigating physical interaction complexities in embedded systems. In: Proceedings of the 5th Malaysian Software Engineering Conference (MySEC2011), pp. 214–219. IEEE. Malaysia (2011)

21. Gulliksen, J.: How do developers meet users?—attitudes and processes in software development. In: Doherty, G., Blandford, A. (eds.) DSVIS 2006. LNCS, vol. 4323, pp. 1–10. Springer, Heidelberg (2007)

22. Biel, B., Grill, T., Gruhn, V.: Exploring the benefits of the combination of a software architecture analysis and a usability evaluation of a mobile application. J. Syst. Soft. 83, 2031–2044 (2010)

23. Ferre, X., Medinilla, N.: How a human-centered approach impacts software development. In: Jacko, J.A. (ed.) HCI 2007. LNCS, vol. 4550, pp. 68–77. Springer, Heidelberg (2007)

24. Folmer, E., Welie, M., Bosch, J.: Bridging patterns: an approach to bridge gaps between SE and HCI. Inform. Soft. Technol. 48, 69–89 (2006)

25. Hussein, I., Mahmud, M., Yeo, A.W.: HCI practices in Malaysia: a reflection of ICT professionals' perspective. In: Proceeding of the International Symposium in Information Technology (ITSim 2010), pp. 1549–1554. IEEE Computer Society, Kuala Lumpur (2010)

26. Joshi, A., Sarda, N.L., Tripathi, S.: Measuring effectiveness of HCI integration in software development processes. J. Syst. Softw. **83**, 2045–2058 (2010)

27. Juristo, N., Ferre, X.: How to integrate usability into the software development process. In: Proceedings of the 28th International Conference on Software Engineering (ICSE 2006), pp. 1079–1080. ACM, Shanghai (2006)

28. Majid, R.A., Noor, N.L., Adnan, W., Mansor, S.: Users' frustration and HCI in the software development life cycle. Int. J. Inf. Process Manag. **2**, 43–48 (2011)

29. Memmel, T., Gundelsweiler, F., Reiterer, H.: CRUISER: a cross-discipline user interface and software engineering lifecycle. In: Jacko, J.A. (ed.) HCI 2007. LNCS, vol. 4550, pp. 174–183. Springer, Heidelberg (2007)

30. Moundalexis, M., Deery, J., Roberts, Kendal.: Integrating human-computer interaction artifacts into system development. In: Kurosu, Masaaki. (ed.) HCD 2009. LNCS, vol. 5619, pp. 284–291. Springer, Heidelberg (2009)

31. Nunes, N.J.: What drives software development: bridging the gap between software and usability engineering. In: Ahmed, S., Jean, V., Desmarais, M.C. (eds.) Human-Centered Software Engineering, Human-Computer Interaction Series, pp. 9–25. Springer, Heidelberg (2009)

32. Vukelja, L., Müller, L., Opwis, K.: Are engineers condemned to design? a survey on software engineering and UI design in Switzerland. In: Baranauskas, Cécilia., Abascal, Julio., Barbosa, S.D.J. (eds.) INTERACT 2007. LNCS, vol. 4663, pp. 555–568. Springer, Heidelberg (2007)

33. Sheridan, J.G., Short, B.W., Van Laerhoven, K.,Villar, N., Kortuem, G.: Exploring cube affordance: towards a classification of non-verbal dynamics of physical interfaces for wearable computing, In: Proceeding of Eurowearable 2003, pp. 113–118. IEEE (2003)

34. Harnad, S.: To cognize is to categorize: cognition is categorization. In: Lefebvre, C., Cohen, H. (eds.) Handbook on Categorization. Elsevier, Amsterdam (2003)

35. Meier, A., Aragon, C., Hurwitz, B., Mujumdar, D., Peffer, T., Perry, D., Pritoni, M.: How people use thermostats in homes: a review. Build. Environ. **46**, 2529–2541 (2011)

36. Carvalho, R., Soares, M.: Ergonomic and usability analysis on a sample of automobile dashboards. Work: A J. Prev. Assess. Rehabil. **41**, 1507–1514 (2012)

37. Blackler, A., Mahar, D., Popovic, V.: Older adults, interface experience and cognitive decline. In: Proceedings of the 22nd Conference of the Computer-Human Interaction Special Interest Group of Australia on Computer-Human Interaction (OZCHI 2010), pp. 172–175 (2010)

38. Blackler, A., Popovic, V., Mahar, D.: Investigating users' intuitive interaction with complex artefacts. Appl. Ergon. **41**, 72–92 (2010)

39. Lawry, S., Popovic, V., Blackler, A.L.: Identifying familiarity in older and younger adults. In: Proceedings of Design Research Society International Conference 2010: Design and Complexity. School of Industrial Design, Université de Montréal, Montréal (2010)

40. Reddy, G.R., Blackler, A., Mahar, D., Vesna, P.: The effects of cognitive ageing on use of complex interfaces. In: Proceedings of the 22nd Conference of the Computer-Human Interaction Special Interest Group of Australia on Computer-Human Interaction (OZCHI 2010), pp. 180–183 (2010)

41. Biswas, P., Langdon, P.: Standardizing user models. In: Stephanidis, C. (ed.) Universal Access in HCI, Part II, HCII 2011. LNCS, vol. 6766, pp. 3–11. Springer, Heidelberg (2011)

42. Langdon, P., Clarkson, J., Robinson, P.: Designing inclusive futures. Univers. Access Inf. Soc. **9**, 191–193 (2010)

43. Persad, U., Langdon, P., Clarkson, P.J.: Investigating the relationships between user capabilities and product demands for older and disabled users. In: Stephanidis, C. (ed.) Universal Access in HCI, Part I, HCII 2011. LNCS, vol. 6765, pp. 110–118. Springer, Heidelberg (2011)

44. Wilkinson, C., Langdon, P., Clarkson, P.J.: Evaluating the design, use and learnability of household products for older individuals. In: Stephanidis, C. (ed.) Universal Access in HCI, Part II, HCII 2011. LNCS, vol. 6766, pp. 250–259. Springer, Heidelberg (2011)

45. Smith, S.L., Mosier, J.N.: Guidelines for designing user interface software. http://hcibib.org/sam/

46. W3C.: Checklist of checkpoints for web content accessibility guidelines 1.0. http://www.w3.org/TR/WAI-WEBCONTENT/full-checklist

47. Hirasawa, N., Ogata, S., Yamada-Kawai, K.: Integration of user interface design process. In: Proceedings of the 11th International Conference on Product Focused Software, pp. 39–42 (2010)

The Cognitive Perception of a Multi-room Music System with Spatial Interaction

Henrik Sørensen, Jesper Kjeldskov[✉], Mikael B. Skov,
and Mathies G. Kristensen

Research Centre for Socio + Interactive Design/Department of Computer
Science, Aalborg University, 9220 Aalborg, Denmark
{hesor,jesper,dubois}@cs.aau.dk

Abstract. In recent years we have seen a growing interest in exploring spatial interaction as a means of interacting with computer systems through what has been labelled "proxemic interaction". In order to explore the potentials and challenges of spatial interaction spanning across separate physical locations, we have developed a multi-room music system and performed a field evaluation of use. The system extends Apple AirPlay to allow spatial interaction with one's music player, for example, adapting an App interface to the current location of the user, and allowing music to follow the user around the house. The prototype was deployed in two households over a three-week period, where data was collected through logging, user-written diaries and interviews. The field evaluation revealed a number of findings related to the cognitive perception of the spaces it was used in, such as importance of a simple interaction, the importance of providing local interaction, the challenge of foreground and background interactions, and challenges in designing interaction with music in discrete zones.

Keywords: Proxemic interaction · Spatial interaction · Multi-room · Music

1 Introduction

Our homes are becoming populated by an increasing number of devices technically capable of interacting with each other. New challenges, therefore, emerge within the field of HCI and work is needed to explore how to take advantage of the possibilities that emerge in the home as a ubiquitous computing (ubicomp) environment.

One way of approaching the interaction design of systems within ubicomp environments is through the perspectives of proxemics and spatial interaction. Hall originally coined the term proxemics as a way of describing interpersonal spatial interaction based on physical measures [11]. In his work he defined discrete zones surrounding us, which are meaningful to the way we interact with each other. The notion of proxemic and spatial interaction used in this chapter is based on the elaboration of Hall's work by Greenberg et al. [10], where they apply proxemics to interaction within ubicomp environments. The purpose is to take advantage of the way proxemics influence how we cognitively perceive, and naturally interact, with each other, and apply this knowledge to interaction within ubicomp environments.

© Springer International Publishing Switzerland 2015
T. Wyeld et al. (Eds.): OzCHI 2013, LNCS 8433, pp. 215–236, 2015.
DOI: 10.1007/978-3-319-16940-8_11

Recent work has explored the possibilities of proxemic and spatial interactions for both work related contexts [10, 13] and leisure contexts [3, 7, 10, 29]. Most studies provide valuable results on what can be called the micro level of proxemic interaction. An example is where content on a large display changes appropriately according to the distance and orientation of the user, as shown in the Proxemic Media Player [10]. On a larger scale, the macro level, spatial interaction may span multiple locations. In many cases this introduces a real cognitive difference and although the macro level some-times consists of several micro level systems, it is important to investigate this type of interaction designs, and see how they are used in real-life contexts.

Music consumption is an area where we see a natural application of proxemic and spatial interactions. Music plays an important role in many people's lives across age, gender and culture. The digitisation of music and advances in mobile and network technologies have opened up for new opportunities in interaction design. Holmquist describes this phenomenon as ubiquitous music [12] and several new advances have recently found its way into the consumer market. An example is the emergence of online music services that contain millions of songs, available through a subscription, making it accessible from several different devices. The growing integration of wireless networks, in our homes, has additionally changed the way we listen to and control music at home. Several music systems, such as Sonos, Bose link, Bang & Olufsen BeoLink and Apple AirPlay, allow us to listen to the same music collection in our entire home, and use mobile devices to control playback. However, on top of the opportunities offered by this new infrastructure there is also an extra layer of com-plexity for interaction design and for research within the cognitive effects of spatial interaction. How do people cognitively perceive interactive systems based on proxe-mics and spatial interaction, especially those spanning across separate physical loca-tions? How does the user choose where to play music? How does the user direct control towards a specific location? The same music could of course play in the entire house simultaneously, but what if different persons want to listen to different songs in dif-ferent rooms? What if the music needs to be louder in a large room and quieter in a small room?

In order to explore some of these questions, we have developed a multi-room music system, called AirPlayer, and performed a field evaluation of use. AirPlayer was designed to hide some of the technical complexity of multi-room music systems through an integration of proxemic and spatial interactions. In our work we focus on how proxemic interactions are used during everyday situations. Due to the importance of the spatial context in which the interaction occurs, findings are based on field evaluations conducted in actual households. In the following we will present related work to proxemic and spatial interaction, spatial cognition, and ubiquitous music. We then describe the AirPlayer system in terms of interaction design and system implementation. We then present the field evaluation, and discuss our findings.

2 Related Work

This section will put our work in relation to existing work to clarify the motivation for exploring proxemic interactions in a music consumption context.

2.1   Proxemics and Spatial Interaction

In the work of Greenberg et al. [10] proxemics and spatial interactions are operationalized in a way suitable to the interaction within ubicomp environments between people, digital devices, and non-digital things. They break proxemics into five specific measurable dimensions: Distance, orientation, movement, identity and location. The dimensions provide discrete and/or continuous measures, which can influence the interaction that takes place. Work originating from this operationalization of proxemics and spatial interaction includes a proximity toolkit for fast prototyping [18], and the use of the sociological constructs, F-formations and micro-mobility, in the design of cross-device interaction [19]. The five dimensions provide a great framework for exploration of proxemics in various ubicomp contexts, and in their work they have furthermore identified six immediate challenges for spatial interaction design [20]. These are, for example, the challenges of directing actions or providing feedback to the user, when the interaction is based on proxemics.

The application of proxemics and spatial interaction in HCI has primarily moved towards different aspects in relation to a central device of focus. Vogel and Balakrishnan [28] have previously specified a set of design principles for public ambient displays. In their work they define a framework, which they refer to as interaction phases. The idea is that the area in front of the device is divided into four discrete phases similar to Hall's proxemic zones [11] surrounding a person. Each phase is determined by the distance to nearby users, and the spatial interaction is described as transitioning from implicit and public to explicit and personal interaction. Recent work has explored similar applications of proxemics, such as public displays [29], whiteboards [13], or tabletops [1, 2]. These studies provide important insight into proxemics in relatively small spaces, and help us understand how proxemics and spatiality can facilitate a different interaction form in such ubicomp environments. There has, however, not been the same focus on proxemics and spatiality on a larger scale.

The idea of exploring proxemic and spatial interactions in larger spaces than a single room, or the immediate area surrounding a display, is of course not entirely new and unexplored. An early, well-known, example is the Active Badge [30]. The Active Badge is aimed at a work context, where employees can be tracked via a wearable badge. The badge contains an infrared beacon from which a sensory network picks up the signal and updates the user position every 15 s Information can for instance be used by receptionists to direct calls to the correct location. The system has successfully been deployed on a large scale at several locations. The UbER-Badge [14] is a different approach to proxemic aware wearable badges. In this case, the badges act as sensor nodes used to facilitate social interaction at large meetings. Each badge contains sensors that can detect other badges or stationary tags. Possible applications include locating other badge-wearers, exchanging contact information wirelessly, and as an interaction device at appropriate locations. The EasyLiving project [4] is an example of proxemic and spatial interaction in the home, with a focus on an architecture that aggregates devices into a coherent user experience. This is accomplished through technologies that track people spatially, and devices and applications binding it together.

Proxemic and spatial interactions on a larger scale, where interaction spans separate rooms of a house, are closely related to the work on indoor positioning, which is a huge research area within ubicomp. Based on measured physical quantity and hardware technology, indoor positioning technologies can be categorised into radio frequency, photonic, sonic waves, mechanical and others [26]. An accurate indoor positioning system is out of the scope of this paper and focus is on the interaction, designed on top of the position system.

2.2   Spatial Cognition

Spatial cognition is a research field dealing with human knowledge and beliefs about the spatial environment around us [21], and working to understand spatial cognition in humans. The field builds on contributions from several disciplines such as Psychology, Geographic Information Science, Human-Computer Interaction, and Cartography. People's spatial perception and cognition is fundamental for our ability to move and navigate through physical space, and for our ability to identify, locate, and track objects and entities in motion [9]. According to recent research, many of the mechanisms that humans make use of in spatial perception and cognition are mainly instinctive, providing us with an ability to store cognitive spatial representations, or cognitive maps [25], for locating themselves, others, things, and directions. In addition to this, because humans can use language for representing space, people are able to create very rich and creative extensions of representations for three-dimensional physical space. This means that physical space plays an important role for humans as a "memorial structure" used to organise memory by, for example, attaching it to specific locations [23].

It is well known that a variety of technologies influence people's spatial cognition, such as the use of the Global Positioning Service (GPS) for navigation, Geographical Information Systems (GIS), and other types of location-based services. Such systems have today become commonplace, and are integrated in hundreds of millions smartphones throughout the world. Nevertheless, it is still a relevant research question to investigate the effect of these systems on people's spatial cognition, and especially how interacting with such systems affect and change people's spatial experience and behaviours, and to investigate how the interaction with such systems is best designed accordingly [21]. How does, for example, the use of location-based services, or proxemics/spatial interaction design, influence people's perception and thinking about their physical surroundings? These questions make it particularly relevant for researchers within the areas of computing and human-computer interaction to apply a spatial cognition perspective on their work, as is exemplified in this chapter.

2.3   Ubiquitous Music

We find the case of music consumption in ubicomp environments particularly interesting as a case for proxemic interaction, as the music itself provides immediate non-visual feedback. This is particularly suitable for a case where location and movement are separated from orientation and in some way distance. Holmquist's description of ubiquitous music [12] embraces music in interaction design for several research directions. The point is that technological advances have changed the way we listen to music. Liikkanen et al. [17] encourages a renewed interest within HCI in music interaction, due to the cultural, social and commercial significance of music consumption, but also because music interaction as a research topic has become less visible.

A radical change in music consumption is the way advances in mobile technologies have enabled us to listen to music on-the-move. In research this has been manifested in a relatively large body of work on mobile music interaction. Recent interesting examples include MusicalHeart, which recommends music based on heart rate [22], and the +++ wearable player [27] that "infects" passers-by with music while jogging. Ubiquitous music is however not restricted to mobile devices. Integration of wireless network infrastructures in our homes open up for similar novel interaction possibilities. The fact that music has become more ubiquitous does not mean that we have stopped listening to music in fixed locations.

Rose [24] reflected on music consumption in the home, back in 2000, and suggested an architecture and interaction design with a central music library and a touch interface. A similar use of music as a case for interaction designs in ubicomp environments is seen in the work of Chang and Kim [6], where a context-aware music playing service is presented. They use a Bluetooth signal to facilitate location recognition, where moving nodes carried by users are detected by a set of fixed nodes. Fixed nodes are then capable of playing music according to the preference of nearby users. What we see is a potential for smarter music systems capable of utilising various devices and a wireless network infrastructure. It is however important also to explore these advances from an HCI perspective.

3 AirPlayer

AirPlayer has been developed to explore proxemics and spatial interaction for a multi-room music system. As a multi-room music system installed throughout the home, it allows users to centralise storage and management of their music collection. Remote control of the music can furthermore be integrated into existing devices. For interaction, AirPlayer makes use of proxemic spatial interaction principles. By making the system aware of the user's location and movement around the house, AirPlayer is able to infer what part of the music system the user wants to interact with, and where music should be played. The goal of this interaction design is to explore the use of proxemic interaction as a form of remote controlling a multi-room music system, and to explore people's cognitive perception of such system in use.

3.1 Multi-room Proxemic Interactions

The proxemic interactions of AirPlayer are implemented as separate spatial features of the system that can be activated independently. The two spatial features, location and movement, are each based on the corresponding dimension from the operationalization of proxemics by Greenberg et al. [10]. In this section, each of these features will be

Fig. 1. The location feature can be activated on the main screen of the music player with the touch of a button.

presented along with a concrete scenario and followed by our interpretation of the spatial dimension in AirPlayer.

Location. The location feature of AirPlayer is activated from the top-left corner of the music player interface as shown in Fig. 1.

The location feature allows the graphical user interface of the smartphone application to adapt to the current location of the user. The following scenario, illustrated in Fig. 2 describes a typical use situation.

> *Alice and Bob enjoy a Sunday evening at home. They are in the living room where Alice is reading a book and Bob is browsing the news on his laptop. Bob has previously used the music browser on the AirPlayer smartphone application to queue a number of songs from their common iTunes library. Because the system knew that he was in the living room, it immediately started playing there as he pressed play. Bob had been up early that morning and goes to the bedroom to get a nap. Bob likes to listen to music as he falls asleep and as a big fan of Bruce Springsteen, he takes out his smartphone to put on his playlist of favourite Bruce Springsteen songs. The smartphone application shows that he is currently in the bedroom and as he selects the playlist, the speakers in the bedroom start playing. The smartphone application shows that it is "The River" playing. Alice is still in the living room listening to the same music as before.*

Fig. 2. When the *location* feature is active, the GUI adapts to the current location of the user.

Greenberg et al. [10] defines the location dimension as a description of the physical context in which the entities reside. Entities in this case can be people, digital devices and non-digital things. They use a particular room as an example of a distinct location, which is also the basis of how it is interpreted in AirPlayer. In AirPlayer the spatial locations are however referred to as zones since the system allows the user to combine separate locations into larger zones. As described in the scenario, the location feature introduces location-awareness of the controlling digital device. When Bob is in the living room with his smartphone, AirPlayer knows this and automatically makes his interactions control the music in the living room. When Bob moves to the bedroom, control is automatically shifted to the bedroom as well, independent of what is playing in the living room. The smartphone application furthermore provides real-time visual feedback to the corresponding zone, showing which zone he is currently in, as well as information about the music playing in that particular zone. Greenberg et al. [10] puts emphasis on the importance of the location dimension, as the meaning applied to the other dimensions can be dependent on the physical context. In AirPlayer this is the case for the movement dimension.

Movement. The movement feature is activated from the top-right corner of the music player interface as shown in Fig. 3.

Fig. 3. The movement feature is activated in the same way as the location feature on the main screen of the music player.

The movement feature tracks the user's movement around between different locations. This information is used to make music follow the user around continuously. The following scenario, illustrated in Fig. 4, explains the use situation.

Charlie is not much of a morning person but music normally helps him get a fresh start of the day. He uses his music system as an alarm clock and at 7:00 AM his stereo starts playing in the bedroom. He grabs his smartphone and goes to the kitchen to get a cup of coffee. The music stops in the bedroom and "follows" him to the kitchen. In the kitchen he suddenly remembers a great song he has not heard for a while. He picks up his smartphone, finds the song in the AirPlayer application, and puts it in the queue. After he finishes his coffee he goes to the bathroom to take a shower and gets ready for work. The music "follows" him to a small speaker in the bathroom and during his shower the song he queued earlier starts playing.

In AirPlayer, movement is interpreted as a discrete measure of changes in the user's spatial location. As illustrated by the scenario, it allows music to follow a person around, by tracking the location of a smartphone. As the person moves in between zones, the system is able to anticipate where the user is going, thereby preparing the music in the zone(s) ahead. Similarly the system is aware of the zone a person is moving away from and stops music playback in that place. As illustrated by the scenario, not only the song playing is transferred, but also queues are transferred between zones as the user moves around.

Fig. 4. When the *movement* feature is active, music follows the user around the house automatically.

3.2   Graphical User Interface

The proxemic spatial interaction in AirPlayer is not a replacement for a graphical user interface (GUI), but a supplement that adds to specific parts of the interaction that otherwise usually complicates the control of a multi-room music system. In addition to the location and movement features, AirPlayer therefore also has a more traditional GUI enabling the user to control other aspects of music playback, such as choosing artists, albums, songs etc. This GUI is available through a smartphone application, which serves as a remote control for the system, and as a display providing visual feedback to the user on their interactions. Using the smartphone application, the user can control music playback in each of the zones through a touch interface, controlling, for example, queuing music on a zone, playing and pausing the music, and adjusting the volume. In order to focus on the proxemics and spatial interactions facilitated by AirPlayer, the GUI functionality is, however, deliberately kept simple, and even when the proxemics and spatial features are turned off, only very basic functionality is available. The GUI consists of three underlying screens: a music player, a queue and a music browser, which we will briefly describe in the following.

Music Player. The music player is the main screen of the application and is also the first screen the user is presented to when starting the application (Fig. 5). The location and movement features can be activated/deactivated by the click of a button placed in the top bar. Once activated, the button will highlight, showing that the feature is active. When movement is active, music follows the user around between zones and naturally so does the control, i.e. the location feature is implicitly active whenever the movement feature is active. The top bar furthermore contains a link button. While the link feature is not directly related to the proxemic interaction, it is used to control the zones, which the proxemic features utilise.

The music playback component is the main part of the music player interface and contains information and controls commonly seen in music players. The interface features a seek bar, controls to start or stop the music, skip to next or previous track, adjust volume, as well as buttons to open the browse and query screens. The music playback component additionally contains a large display of information regarding the

Fig. 5. The music player is the point of entry and primary interaction screen of the application.

music currently playing. The bottom of the music player is reserved for the zone indicator, which displays the name of the current zone being controlled. A swipe gesture to the right or left will manually cycle through the configured zones. When the location feature is active, it will automatically change the zone according to the location of the user.

Queue. The queue screen, shown in Fig. 6, presents the current queue of songs for the active zone, as well as controls to manage the queue. The primary part of the screen is the queue of songs presented in a scrollable list. The clear button in the top right corner removes all songs queued in the active zone, and stops playback. Controls in the bottom can be used to play a song, rearrange the queue, or remove songs.

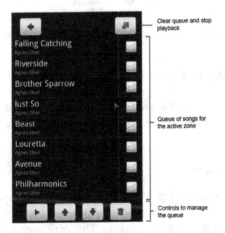

Fig. 6. The queue for the active zone can be accessed and managed through the user interface.

Music Browser. The music browser, shown in Fig. 7, has a top bar with a back button, an add button, and an add-all button. If the user presses the add button, all selected songs are added to the queue of the active zone. The primary part of the music browser, is a list of items from the music library, i.e. playlists, albums, artists and songs. The list is sorted alphabetically and the user can scroll using swipe gestures. The bottom bar of the interface contains a tabbed interface for the user to browse through the music library. For example, when the user selects the Albums tab, a list of available albums from the iTunes library is presented. From here, the user can select an album, and all songs on the album will be displayed on the list. From the song level, the user can also select a number of songs to add to the zone queue, or simply add all songs from the album to the queue. In this way, the user can browse through the iTunes library from his smartphone, and add the desired songs to the queue of currently playing songs.

Fig. 7. On the song level of the music browser, songs can be added to the queue.

4   System Implementation

In this section we provide details about the implementation of AirPlayer, including system architecture and details about the location estimation and server application.

4.1   System Architecture

The AirPlayer prototype system platform is build on top of Apple's AirPlay, which allows streaming of media content between Apple products, and Apple compatible products, using a regular wireless network. The AirPlayer infrastructure relies on three types of devices: (1) a smartphone which hosts the AirPlayer client application,

(2) a Mac Mini server that hosts the AirPlayer server application, the music collection, and iTunes, and (3) a number of AirPort Express wireless network base stations, connected to either active speakers or a hi-fi stereo system.

The AirPlayer system architecture is depicted in Fig. 8, which shows relations between components. The server application is installed on a Mac Mini server and is basically a remote controlled music player capable of streaming to AirPort Express stations. It gets the music data directly from the music collection residing on the server. Apple's music player iTunes is also installed on the server, but only serves the purpose of accessing meta-data from the music library, and providing an interface where the user can manage the music collection and playlists. In addition to receiving the music stream, the AirPort Express wireless base stations also provide location measurements handled by the smartphone application. The smartphone application communicates directly with the server application to issue commands like "skip to next song" or "move music stream to another AirPort Express". It furthermore receives data from the server, such as meta-data about the currently playing song, which can be presented to the user. The user can control the system using a smartphone from anywhere in the house directly through touch input or indirectly by activating the proxemic interaction features.

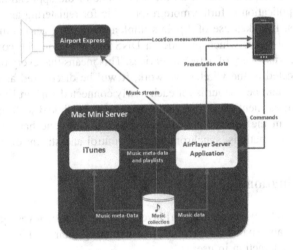

Fig. 8. The system architecture of AirPlayer consists of three types of inter-connected devices.

4.2   Location Estimation

The use of location awareness is a central part of AirPlayer. Both the location and movement features are reliant on knowledge about the user's current spatial position within the environment. In AirPlayer this is achieved through a simple comparison of received signal strength indicator (RSSI) measurements, handled by the smartphone application. For the prototype application, only coarse-grained information regarding the user's position is required, i.e. in which zone the user is currently located and whether his location has changed over time. By using results from Wi-Fi RSSI measurements from the AirPort Express devices, the relative location of the user can be estimated, thus creating the basis for the implementation of the proxemic interaction.

AirPlayer is a self-positioning system [8], meaning that the positioning receiver is responsible for doing appropriate measurements and uses these to position itself. In AirPlayer the mobile application is responsible for collecting the RSSI measurements, from each available AirPort Express, and use them to determine which zone it is currently in. Having the smartphone application as a self-positioning receiver makes it less dependent on changes in the infrastructure and therefore does not need to be aware of details about the setup such as the number and location of available AirPort Express stations.

4.3   Server Application

The server application is the backend of the system and is basically a remote controlled music player capable of streaming wirelessly to speakers connected to an AirPort Express. It contains information about the user's iTunes music collection including playlists. Actual management of the music, like editing playlists or reorganising the music, is handled directly in iTunes. This provides a familiar interface for the user and results in a simpler implementation on the server side of the application.

The server application is furthermore responsible for registering and managing the configured zones. It makes use of Apple's implementation of zero configuration networking, called Bonjour, which provides a DNS based Service Discovery API that enables automatic discovery of network services. This means that every time an AirPort Express is connected to the wireless network, it will be discovered automatically by AirPlayer. The default behaviour is for each newly connected AirPort Express to create its own zone, but as mentioned, individual zones can be linked to form larger ones. This is managed in the server application by letting one zone handle a number of AirPort Express connections, synchronizing the control and stream of music.

5   Field Evaluation

We conducted a field evaluation of AirPlayer with the goal of investigating people's use of proxemics and spatial interactions, and exploring their cognitive perception of such system and interaction in use.

5.1   Method

The field evaluation spanned three weeks where AirPlayer was installed and integrated with participants' existing multi-room music system at home. The participants were introduced to the system in the beginning of the evaluation and then used it in their everyday lives for the duration of the study. Participants were asked to note their thoughts about the system in a diary. After the three weeks a semi-structured interview was conducted, following the guidelines of Lazar et al. [16]. Entries from the diaries were included as a basis for a conversation regarding the system. Interviews were conducted in the homes of the participants, to provide a comfortable environment and to let the participants talk about their experiences in the place where interaction took place.

Two interviewers were present at each session. One would specifically be responsible for taking notes including information about esoteric remarks, visual references etc. Interviews were furthermore recorded.

5.2 Participants

Two households, which will be referred to as A, and B, participated in the field evaluation. Both had a multi-room music system installed prior to the evaluation and were therefore familiar with the basic concept and use of such systems. The participants received a small token of appreciation for participating in the evaluation.

Household A. The first participating household had two residents. A woman aged 47 (A1) living with her 16 year old son (A2) on a small farm. The music player currently installed in their home was a Sonos system. The woman had four years of experience with the system and her son one year. Three AirPort Express stations were installed in the locations where the participants usually listened to music using their current system as illustrated in Fig. 9.

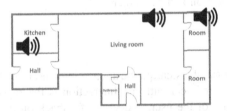

Fig. 9. Floor plan of household A illustrating placement of AirPort Express stations. The floor plan only includes the part of the farm where AirPlayer was installed.

Household B. The second participant was a 28 year old man living alone in an apartment (B1). He already had a setup consisting of Apple products and used iTunes to play music. AirPlayer could therefore easily connect to the existing system. An additional AirPort Express was added to get more than two locations. The floor plan is shown in Fig. 10.

Fig. 10. Floor plan of household B illustrating placement of AirPort Express stations.

6 Findings

Basic usage was logged by the system itself, which will be presented to give an overview of the foundation for the qualitative findings.

Over the three-week period household A used the system for 70 h in total, averaging 3.3 h per day. The system would however be used more in the first half of the evaluation period. During the 70 h of use, the location and movement features were active for 57 % and 23 % respectively. The usage in household B had a more even distribution with a total of 43 h, averaging 2.0 h per day. The location feature was active 47 % of the time and the movement feature 39 %. Reasons for the relatively low percentages of time the features were active can partly be found in the findings presented later. There are however also a few low-practical explanations. One is the location feature being implicitly active when the movement feature is active, but not counted in the statistics if it is not explicitly activated in the system. Another is the fact that participants would turn features on one at a time to explore this particular feature.

From this usage, participants were able to continuously report experiences in their diary and conclusively report on their use of the system during the interview. The following sections describe our findings, structured by the two different proxemic interaction features, Location and Movement.

6.1 Location

The location feature allows the smartphone application's user interface to automatically adapt to the zone the user is currently in. Interaction is thereby directed towards the current spatially location of the user, and visual feedback on the smartphone reflects what is playing in that particular location. From the interviews two main findings in relation to this type of spatial interaction design were identified, specifically regarding simple interaction and local interaction.

Simple Interaction. The purpose of the proxemics features of AirPlayer is to hide the complexity of interaction with the music system by introducing a layer of spatial interaction. Both households described the location feature a very simple way of interacting "spatially" with the system. They perceived this form of interaction with the music player as intuitive, but also as very transparent. In fact, A1 expressed that she was not even always consciously aware that she was using it:

> To me it had to be fantastic, because I used it and I didn't even think about whether or not the feature was enabled. I actually thought it was connected with the other [feature] where music followed me from room to room. It worked when I adjusted the volume and other things. But I had not thought about the fact that I used it, but I really did. (A1)

The fact that she was not always consciously aware that she had enabled the location feature, despite of not having this function available in her existing Sonos system, indicates that the feature was very subtle, and that automatically adapting the interface of a smartphone application to the user's immediate spatial surroundings matches the users' spatial cognition in relation to that place and that type of interaction. This is especially evident since the location feature was in fact active 57 % of the time in household A, and also implicitly active when the movement feature was enabled as well. Looking back at her interaction s with the system, the user of household A understood immediately how the location feature worked, and that interacting spatially in her home, by moving from one room to another had a direct effect on the user interface. She explicitly thought of it as a "simple" form of interaction to be able to use her smartphone to control the music in the room she was currently in – just as she was doing when the movement feature was active. In household B, the participant similarly experienced the location feature to be simple. Compared to household A, however, he was consciously aware that he was using this particular feature:

> As I said earlier, I think that a relatively simple feature as this one is extremely good. The fact that I do not have to find the room that I am about to play music in, makes it easy to utilise the mobile phone to control the music according to your current location. (B1)

Another point where the need for simple interaction was visible was not in the added proxemic features but instead in features missing. Participants from both households commented that they were missing the radio or Internet radio functionality of the systems they were used to. B1 also mentioned that he would have liked a larger queue size. Both of these comments indicate a need for simple interaction where it is not up to the user to constantly control the music in several locations.

Local Interaction. What we experienced from the field evaluation was that the location feature, which limits interaction to the immediate spatial surroundings of the user, was greatly appreciated by the participants for its spatially localised interaction. By enabling the location feature, they found it easy to use the application as a localised remote control for the music player, as the application was aware of its current location and could therefore easily manage the music in the given zone. Both households used the location feature during the majority of the test period. As the participant from household B expresses:

> I would estimate that I just used it. I just used it most of the time. Actually, I have had no real need to control another room other than the one I was present in. (B1)

This is also supported by the previous comments from A1 who used the location functionality without being consciously aware of it, because she experienced that it just "naturally responded nearby" whenever she interacted with it.

6.2   Movement

The second proxemic dimension, movement, was also implemented as a part of the interaction design of AirPlayer. When active, the movement feature allows music to follow the user around the house additionally moving the queue and control of music as well. From the interviews, two main findings in relation to this type of spatial interaction design were identified, specifically regarding foreground and background interaction and music in discrete locations.

Foreground and Background Interactions. None of the participants initially expressed any uncertainties regarding the concept or functionality of AirPlayer. When asked to describe the system's procemic and spatial interaction features, using their own words, they were all able to give a brief and correct overall description of how it actually worked, suggesting that the implemented interaction design matched the users' spatial cognition in relation to that specific physical place very well. However, B1 also elaborated:

> *Of course, I had to get used to how it worked the first time I used it, but I do not think that we had any doubts of how it was supposed to work. (B1)*

Both households had a good overall perception of how the different features worked spatially. Despite that, however, we did find examples where the details of the Air-Player system's behaviour was not clear:

> *I would have thought that the music stopped when I went upstairs to bed. It didn't. It stayed in the last room I left – That puzzled me. (A1)*

Because the music "followed" A1 into rooms where a speaker was present, as in started playing when entering a room, she also expected the music to stop playing when she left one of those rooms - or at least when she left the floor of the house where that particular room was located. This was not the case in the current implementation of AirPlayer, leading to some surprise and uncertainty. From a spatial cognition perspective, the issue at play here is simply that the system does not consistently behave naturally, but has different behaviour depending of whether the user enters or leaves a room. If the spatial experience of music playback "following" the user was to be realised fully, music should, of course, also stop playing when leaving a particular room. Alternatively, the spatial experience created would be one of "leaving a trail" of music rather than the desired one of "taking it with you". In relation to this issue, B1 stated that he actually thought that the music had stopped playing in the rooms he had been to and then left. But because he had left the room, he was unaware that the music was in fact still playing.

Music in Discrete Zones. Interesting findings also emerged regarding the relationship between music and the spatial subdivision of the house into discrete zones. In particular, both families found that the movement feature worked well. As one of the participants from household B stated:

> *We have used it a lot, partly because we wanted to test the system, but also because we found it to be clever and fun to use. The thing about having the music following you is nice when you walk around at home in your own thoughts. (B1)*

Household A shared this opinion towards using the feature; however they also experienced a slightly unintended behaviour. When asking how the feature behaved, they explained that the music should overlap for longer periods than it currently did:

> *It could have waited 15 s before it stopped the room you were leaving, and then start the music immediately as you walk into the next room. I think more overlap would have worked wonders (A1)*

As participant A1 stated, she would have liked AirPlayer to have a timer-based threshold instead of having one purely based on distance between two zones and the smartphone. The result of the behaviour was that she would turn off the feature while cleaning the house, as she would move frequently between different rooms.

At times, the participants also experienced that the music would not be entirely synchronised, which were obvious when standing in the middle of two different zones with the movement feature active. The result was a slightly asynchronous playback of the same song:

> As I said earlier, there were a few incidents where the music was not entirely synchronised, but we learned that we could fix it by changing the track being played and then switching back (B1)

The asynchronous behaviour was caused by technical limitations discovered in the preliminary study when the framework was developed.

7 Discussion

Several systems previously presented within research on proxemics and spatial interactions are based on a central visual focal point, like a large public display, where interaction takes place in the immediate surroundings. Complementing this research, the AirPlayer prototype has been developed to explore proxemic interactions in environments that span separate locations, and where the primary medium is not visual but auditory. Our focus has furthermore been on exploring specific spatial dimensions of proxemic interaction, namely location and movement, and to understand better the effects of such spatial interaction in real world use. In this section we will discuss further the cognitive perception of the proxemics and spatial interaction with AirPlayer in people's homes.

7.1 Simple Interaction

One of the things that stood out from our field evaluation was that proxemics and spatial interaction provided a means to facilitate simple interaction with the multi-room music system. The point of simplicity in this case is not only that the interaction is simple to understand and use, but also that the proxemic and spatial features reduce the complexity of what is in fact a relatively complicated setup of distributed speakers and wireless media players. The fact that one of the participants (A1) was unaware that she had used the location feature, but still found it useful when reflecting on it in hindsight, points out a particular positive potential of proxemics and spatial interactions when they are designed and implemented well. Interaction can be very transparent to the user as long as it feels natural, and letting features become invisible to the user can even be a success criterion of the design. Our findings about the use of the location feature did not indicate importance of being informed every time control shifted from one spatial zone to another, or that people needed advanced features to feel more in control of the system at their spatial location. One reason can be that the metaphor of a single music system in each room simply becomes stronger with the location feature active, and that it then just feels natural to use such a system as a remote control for the speakers in people's immediate spatial surroundings.

7.2   Local Interaction

The main differences between multi-room music systems and traditional music players is that the music source and the control can be centralised when playing music in different spatial locations of the home. What the location feature of AirPlayer does is really to automatically limit the control to the room the user is currently in. During the field evaluation of AirPlayer, this feature was well received, and it was found to be very intuitive to control music "locally" in spatial vicinity of the user. While integration of wireless networks in our homes provide great opportunities for remotely controlling everything from everywhere, our findings point toward positive aspects of doing the opposite, and use spatial positioning to localise control to only those systems present in the users immediate vicinity. By doing so, we found that people very quickly perceive control systems, for example an application on a smartphone, as extensions of the interaction possibilities in their spatial location, and very quickly understand that the system changes when they move between physical locations. This observation confirms previous work on context-awareness, which has, for example, described spatially dependent mobile systems as dynamic indexical signs, which people are generally highly capable of interpreting [14]. This finding also contributes to the argument that discrete zones with localised interaction can be powerful for the user in order to understand where interactions are directed [20].

7.3   Foreground and Background Interactions

Buxton talks about foreground and background interactions within HCI [5]. Where the graphical user interface of the smartphone application facilitates foreground interaction, proxemic and spatial interactions facilitate background interactions. Making the user understand the intended design of background interactions can be rather challenging exactly because they are supposed to be more discrete. What the use of the movement feature in AirPlayer showed us was that although the overall concept was easy to understand, the functional details about the interaction were difficult to communicate to the user. One participant knew that the movement feature allowed music to follow her around and therefore also expected it to stop when she went to a room where music could not be played. As a consequence of making an effort to simplify the interaction, in our case through proxemic and spatial interactions, challenges in making details about this interaction design apparent to the user were unintentionally introduced. In our specific case this was a particularly problematic challenge because the activity of listening to music is already typically something that takes place in the background.

7.4   Music in Discrete Zones

Certain issues and unexpected behaviour were experienced by the participants in relation to the movement feature. In response to the way music moved between zones, A1 preferred not to use the movement feature while cleaning the house, as she would be moving frequently between rooms. The system behaviour she experienced might be due to the specific layout of the house, but it does raise an interesting issue in relation to the combination of spatial interaction and auditory output. One of the reasons for the annoyance could be because the speakers, often placed in opposite sides of adjacent rooms, created the experience of music jumping from one side of the house to the other, instead of creating a seamless experience. Another reason is possibly that while the location system is divided into discrete zones, music is not restricted to be audible within a specific zone but easily travels across zones, even through walls. Because of this, when standing between two zones, users also sometimes experienced that playback from different sets of speakers was slightly out of sync. While this may be avoided by implementing the system slightly differently in order to achieve better synchronisation, the fact that sound travels differently to visual media presents some fundamental challenges to proxemics and spatial interaction with such systems.

7.5   Cognitive Effects of Spatial Interaction with AirPlayer

Our study has showed that deploying an auditory media system with proxemic and spatial interaction into people's homes has an effect on how people cognitively experience the relationship between such system and the physical space where it is being used. It is clear that the layout of the physical space influences people's experience of the system, and their expectations about how it should behave. But it is also clear that the use of the system conversely influences people's experience of the physical space it is being used in.

Starting with the effect of physical space on expected system behaviour, we saw very clearly that people readily experienced the localised user interface, and the music being played in a room, as just another natural property of a particular space, like its physical shape, colour, lighting, etc. Hence, it was also readily expected that the user interface would change when moving to another spatial location, and that when in another spatial location one would be able to control the music playing there. This leads us to conclude that there is a significant cognitive effect of people's placed-ness in the physical world on their perception of interactive digital systems that respond to proxemic and spatial relationships and changes in their physical surroundings.

In terms of the effect of systems with proxemic and spatial interaction on the experience of the physical space it is being used in, we saw a change in people's spatial experience of their homes in the sense that some physically separated locations were suddenly less separated because sound triggered by the interaction with the system would now flow between rooms. On the positive side this meant that previously disconnected physical spaces could now be experienced as connected ones, contributing to an experience of wholeness when moving through the house. On the negative side, however, it also meant that those spaces that people wanted to keep separated could be more difficult to experience as such. This was specifically the case when separate music sources sometimes interfered with each other across rooms, but it was also the case when people experienced that they were unable to go into a silent room, for example to answer a phone call, without the music "chasing" them. This exemplifies an unwanted cognitive effect of perceiving the house, rather than the inhabitants, being in charge of how the physical space is used, and how it can be negotiated. In relation to this, we find it very important that systems with proxemic and spatial interaction do not take control away from the user.

7.6 Technical Limitations of the Spatial Tracking

We are aware of the limitation imposed by the direct use of RSSI values as indication of physical distance, due to uncontrollable external influences. A more accurate indoor positioning system is however out of the scope of this paper and the focus has been on the concept of proxemic interactions in real-life contexts. What is interesting to note is that even though participants were asked about the usefulness of the proxemic interaction features, no comments were directly addressing problems with accuracy of the location estimation, despite the simple implementation. Other issues have surfaced during the interviews, but an overall perception of proxemic awareness and satisfaction with the features have been experienced by the participants.

8 Conclusion

We have explored people's experiences of spatial interaction with a multi-room music system by developing a functional prototype and deploying it in actual households. This has enabled us to investigate the effects of two specific dimensions of proxemics and spatial interaction, namely location and movement. The study yielded a number of specific findings related to simple interaction, local interaction, foreground and background interaction and music in discrete zones. The findings suggest that proxemics and spatial interactions have a great potential for "hiding" parts of the user interaction with a complex system and thereby making it simpler for people to use. However, we also found that it is challenging to simplify background interactions in a way where details about the functionality and possibilities for interaction are still apparent to the user. We also identified challenges in relation to combining the specific medium of music with spatial interactions, specifically related to the fact that audio may travel across the physical boundaries delimiting our spatial surroundings.

Acknowledgements. A special thanks goes to the field evaluation participants, and to Helle Hyllested Larsen for her work on illustrations.

References

1. Ackad, C.J., Clayphan, A., Maldonado, R.M., Kay, J.: Seamless and continuous user identification for interactive tabletops using personal device handshaking and body tracking. In: Proceedings of CHI EA 2012, pp. 1775–1780. ACM Press (2012)
2. Annett, M., Grossman, T., Wigdor, D., Fitzmaurice, G.: Medusa: a proximity-aware multitouch tabletop. In: Proceedings of UIST 2011, pp. 337–346. ACM Press (2011)
3. Ballendat, T., Marquardt, N., Greenberg, S.: Proxemic interaction: designing for a proximity and orientation-aware environment. In: Proceedings of ITS 2010, pp. 121–130. ACM Press (2010)
4. Brumitt, B., Meyers, B., Krumm, J., Kern, A., Shafer, S.: Easyliving: technologies for intelligent environments. In: Thomas, P., Gellersen, H.W. (eds.) HUC 2000. LNCS, vol. 1927, pp. 12–29. Springer, Heidelberg (2000)

5. Buxton, B.: Integrating the periphery and context: a new model of telematics. In: Proceedings of GI 1995, pp. 239–246. Canadian Information Processing Society (1995)
6. Chang, J.-W., Kim, Y.-K.: Context-aware application system for music playing services. In: Gabrys, B., Howlett, R.J., Jain, L.C. (eds.) KES 2006. LNCS (LNAI), vol. 4253, pp. 76–83. Springer, Heidelberg (2006)
7. Clark, A., Dünser, A., Billinghurst, M., Piumsomboon, T., Altimira, D.: Seamless interaction in space. In: Proceedings of OZCHI 2011, pp. 88–97. ACM Press (2011)
8. Drane, C., Macnaughton, M., Scott, C.: Positioning GSM telephones. IEEE Commun. Mag. 36(4), 46–54, 59 (1998)
9. Evans, V., Chilton, P.: Language, Cognition and Space: The State of the Art and New Directions. Equinox Publishing, London (2010)
10. Greenberg, S., Marquardt, N., Ballendat, T., Diaz-Marino, R., Wang, M.: Proxemic interactions: the new ubicomp? Interactions 18(1), 42–50 (2011). ACM Press
11. Hall, E.T.: The Hidden Dimension. Doubleday, Garden City (1966)
12. Holmquist, L.E.: Ubiquitous music. Interactions 12(4), 71 (2005)
13. Ju, W., Lee, B.A., Klemmer, S.R.: Range: exploring implicit interaction through electronic whiteboard design. In: Proceedings of CSCW 2008, pp. 17–26. ACM Press (2008)
14. Kjeldskov J., Paay J.: Indexicality: understanding mobile human-computer interaction in context. ACM TOCHI 17(4), 14 pp. (2010)
15. Laibowitz, M., Gips, J., Aylward, R., Pentland, A., Paradiso, J.A.: A sensor network for social dynamics. In: Proceedings of ISPN 2006, pp. 483–491. ACM Press (2006)
16. Lazar, J., Heidi, J., Hochheiser, H.: Research Methods in Human-Computer Interaction. Wiley, Chichester (2010)
17. Liikkanen, L., Amos, C., Cunningham, S.J., Downie, J.S., McDonald, D.: Music interaction research in HCI: let's get the band back together. In: Proceedings of CHI EA 2012, pp. 1119–1122. ACM Press (2012)
18. Marquardt, N., Diaz-Marino, R., Boring, S., Greenberg, S.: The proximity toolkit: prototyping proxemic interactions in ubiquitous computing ecologies. In: Proceedings of UIST 2011, pp. 315–326. ACM Press (2011)
19. Marquardt, N., Greenberg, S.: Informing the design of proxemic interactions. IEEE Pervasive Computing 11(2), 14–23 (2012)
20. Marquardt, N., Hinckley, K., Greenberg, S.: Cross-device interaction via micro-mobility and f-formations. In: Proceedings of UIST 2012, pp. 13–22. ACM Press (2012)
21. Montello, D.R.: Spatial cognition. In: Smelser, N.J., Baltes, P.B. (eds.) International Encyclopedia of the Social & Behavioral Sciences, pp. 14771–14775. Pergamon Press, Oxford (2001)
22. Nirjon, S., Dickerson, R.F., Li, Q., Asare, P., Stankovic, J.A., Hong, D., Zhang, B., Jiang, X., Shen, G., Zhao, F.: MusicalHeart: a hearty way of listening to music. In: Proceedings of SenSys 2012, pp. 43–56. ACM press (2012)
23. Nova, N.: Socio-cognitive functions of space in collaborative settings: a literature review about Space, Cognition and Collaboration. CRAFT Research Report No. 1 (2003)
24. Rose, M.: Music in the home: interfaces for music appliances. Pers. Technol. 4(1), 45–53 (2000). Springer
25. Tolman, E.C.: Cognitive maps in rats and men. Psychol. Rev. 55, 189–208 (1948)
26. Torres-Solis, J., Falk, T.H., Chau, T.: A review of indoor localization technologies: towards navigational assistance for topographical disorientation. In: Jesus, F., Molina, V., (eds.) Ambient Intelligence (2010)
27. Trotto, A., Tittarelli, M.: Musical viruses for graceful seduction. In: Proceedings of NordiCHI 2012, pp. 731–735. ACM Press (2012)

28. Vogel, D., Balakrishnan, R.: Interactive public ambient displays: transitioning from implicit to explicit, public to personal, interaction with multiple users. In: Proceedings of UIST 2004, pp. 137–146. ACM Press (2004)
29. Wang, M., Boring, S., Greenberg, S.: Proxemic peddler: a public advertising display that captures and preserves the attention of a passerby. In: Proceedings of PerDis 2012, ACM Press, Article no. 3 (2012)
30. Want, R., Hopper, A., Falcão, V., Gibbons, J.: The active badge location system. ACM Trans. Inf. Syst. **10**(1), 91–102 (1992)

Glossary

Author Index

Printed in the United States
By Bookmasters